THE RISE OF THE EVOLUTION FRAUD

San Diego Christian College
Library
Santee, CA

Charles Darwin in 1855 and 1869

213
B784r

THE RISE OF THE EVOLUTION FRAUD

(An exposure of its roots)

by

M. BOWDEN

CREATION·LIFE
PUBLISHERS
P.O. Box 15666
San Diego, California 92115

Copyright © M. Bowden 1982
All rights reserved. No part of this publication
may be reproduced, stored in a retrieval system, or
transmitted, in any form or by any means, electronic,
mechanical, photocopying, recording or otherwise,
without the prior permission of the copyright owner.
Short extracts may however be quoted on the sole condition
that the title of this work, the authors name and the
publisher is referred to with every passage quoted.

ISBN 0-89051-085-7

First Edition 1982

Co-published
by
M. Bowden
Bromley, Kent, England
and
Creation-Life Publishers
San Diego, California

The text was written on a Commodore 'PET'™ Computer using a
Wordpro III wordprocessor program. Phototypeset direct from
disks by REPRODESIGN, 131 Market Street, Chorley, Lancashire.

FOREWORD

Having read Malcolm Bowden's outstanding book *Ape-Men: Fact or Fallacy?* several years ago, I was very pleasantly surprised to meet him recently at a Christian workers' conference at the Metropolitan Tabernacle in London, famous as Spurgeon's great church back in the nineteenth century. I had been speaking about the hidden background and modern impact of evolutionism, commenting on the need for some creationist scholar to make a thorough and original study of these two important subjects. Malcolm Bowden, after introducing himself, announced that he was then completing just such a manuscript!

This book is the product of his studies and is, indeed, a unique treatment of the remarkable phenomenon of evolutionism. In evolution we have a supposedly scientific world view for which there is no real scientific evidence whatever. Furthermore, it has spawned a vast complex of harmful philosophies and methodologies in the world during the past two centuries. Yet it is accepted and promoted, dogmatically and evangelistically, in the schools and universitites of every nation in the world! Such a phenomenon surely warrants a more critical and incisive explanation than is now available in run-of-the-mill liberal treatments of the history and philosophy of evolutionary thought.

The Rise of the Evolution Fraud does, indeed, throw much new light on this remarkable epoch of history. Exploring the personal and environmental backgrounds of Darwin and his contemporaries, Malcolm Bowden has placed the whole subject in a new perspective. Especially significant are the new insights into Darwin's own psychological conflicts and the intriguing role played by the lawyer/geologist Charles Lyell in promulgating the rise of evolutionism. Bowden also adds a fascinating appendix on the question of Darwin's possible conversion to Christianity in the last weeks of his life. Finally, he provides an incisive critique of the logical basis of evolution, demonstrating its fatal scientific and philosophical weaknesses, as well as its devastatingly harmful impact upon society.

This book is bound to be controversial and will, no doubt, incur the vengeful wrath of the evolutionary establishment. In a sense it is still only an exploratory treatment, but its very publication should establish the need and stimulate the development of much further research in these highly important, but still dimly appreciated and understood, facets of modern evolutionary humanism. Evolutionism is far more than a mere theory of biology. It is a complete world view, with its roots extending far back through the entire history of man's rebellion against his Maker and with its deadly fruits entering the bloodstream of every field of human thought and activity today. It is vital that men and women somehow acquire a real understanding of its true character.

Malcolm Bowden has produced a book which can make a vital contribution toward that end. I am pleased and honored to commend it to the thoughtful study of readers everywhere.

Dr. Henry M. Morris
March 1982

Acknowledgements

I must record my grateful thanks to several friends,
in particular Dr. D.T. Rosevear, who read the manuscript and
made a number of valuable suggestions.

Dedication
**This book is sincerely dedicated to my wife –
who bore the brunt.**

Notes
1. All emphasis in italics in quoted passages are by the
author of this book unless otherwise noted.

2. Reference numbers of publications are given in the bibliography.

THE RISE OF THE EVOLUTION FRAUD

CONTENTS

CONTENTS

SECTION IV

THE SECRET AIM – ACHIEVED

APPENDICES

LIST OF ILLUSTRATIONS

List of illustrations to Appendices
(Following page 209)

Ad Majorem Dei Gloriam

INTRODUCTION

It is commonly thought that Charles Darwin was responsible for conceiving the basic idea of "Evolution". Those who study the matter further usually agree that Darwin first had doubts about the "fixity of species" as a result of examining the variety of finches on the Galapagos Islands. It is then thought that it was this basic idea which he later developed and expounded in his book *The Origin of Species* which was first published in 1859. This idea has been so well publicized over a considerable period of time that it is now the most popular interpretation of these events. However, like many other popular and highly-publicized notions, it is quite false.

The real truth is that at no time during his voyage did he seriously doubt that species were created and could only vary within certain limits. Similarly, throughout the whole of his voyage on the *Beagle*, he was a *creationist*, and it was not until almost a year after his return that he first had any thoughts on the subject of evolution!

This information, and very much more besides, is to be found in *Darwin and the Darwinian Revolution* by Gertrude Himmelfarb [1]. It is here that I must pay tribute to this most thoroughly researched biography of Darwin's life and his relationship with his colleagues and friends for it is this work which mentioned aspects that were well worthy of further investigation. In most biographies of Darwin, one sees the same facts repeated in each of them, and there is a strong suggestion that later works simply copy earlier versions. When book after book repeats the same errors, the ordinary researcher can be forgiven for concluding he is obtaining well corroborated information. It is only when a biographer as painstaking and meticulous as Himmelfarb takes the trouble to refer not just to previous biographies or even the published volumes of Darwin's letters but to Darwin's original notebooks and correspondence, that a strikingly different picture emerges. Perhaps even more important is her very objective assessment. She is clearly not a creationist but finds the arguments of the evoltionists far from convincing. Her approach is refreshingly free from the usual adulation and hero-worship that invariably surrounds anything to do with Darwin or his colleagues. Indeed, her close examination of their popular images leaves them with little credibility.

I am indebted to this work for some of the detailed information of Darwin's life and others, particularly her quotations from unpublished letters. I would recommend it to all who wish to examine the background to the rise of evolution and the many personalities involved. It is singularly unfortunate that this biography, which exposes so many errors surrounding the rise of the theory, should be out of print, with no plans by the publishers to reprint this valuable work. I have therefore quoted rather more items than I would otherwise have done in view of the difficulty of obtaining this book.

I must also record my debt to two others. The first is R.E.D. Clark for his book *Darwin, Before and After* which contains a number of valuable insights. The second is Dr. Arthur Jones, whose personal help and unpublished monographs were of great value when dealing with the subject of the philosophical aspects of science and evolution and the subject of cortical inheritance.

The reader may be interested to know the background to the writing of my first book and the present work, for it is more the story of the book which did not get written!

It was at the first Creation Conference at Swanwick in 1974 that I considered the possibility of writing a book on *The Scientific Evidence against Evolution*. Several months later I drafted the chapter headings and began writing on the section dealing with the Ape-men fossils. I found however that the deeper I probed the evidence, the more surprising the facts which revealed themselves. Eventually, there was sufficient material on this one topic alone and I wrote my first book *Ape-men — Fact or Fallacy?*. Three years later, returning to the original book planned, I wrote virtually the whole of the section on the geological evidence, in the course of which I began to examine the life of Charles Lyell.

It was only then that I realised the important influence which he had upon the rise of the theory of evolution. It is generally thought that his main contribution was in providing the vast periods of geological time which Darwins theory required. Close examination of his life and letters however showed that his real purpose was, by means of a long term scheme, to overthrow the "Mosaic account" - i.e., the Genesis account of the Creation,- even before Darwin had set sail on the *Beagle*. A study of his very great influence on the course of Darwin's work, together with the activities of Huxley and the secretive "X Club" finally prompted me to produce the present volume.

All too frequently today, biographies appear which achieve some notoriety simply because they publicise some of the less attractive qualities of a famous person. Doubtless many will wish to place this book in the same category. I would however claim that my intentions are quite different. When biographies are written which denigrate a particular personality, this is often done by exposing their personal weaknesses and the scandals of their private lives, whilst criticisms of the major achievements which made them famous may be few. It is quite the opposite in this case. As far as I could ascertain, every single one of the men referred to in this work, in both their public and family conduct, led perfectly respectable lives, were men of their word, excellent husbands and pillars of society.

When however we consider the motivation and covert activities of a small group of men intent on persuading the public to accept a theory which was not only demonstrably false even in their own day, but also to have appalling consequences on moral attitudes throughout the world, then a work which publicises their aims and activities is surely long overdue.

SECTION I

EARLY THEORIES

CHAPTER 1

ANCIENT IDEAS

Darwin consistently claimed that he originated the Theory of Evolution, for in the early editions of his *Origins* he wrote of it as 'his' theory, and in his autobiography he denied that the subject was "in the air", for he had "occasionally sounded out a few naturalists, and never happened to come across a single one who seemed to doubt about the permanence of species" [2p89]. It is true that the vast majority of the scientists of his day did accept the permanence of species for they had no evidence to the contrary. But it is also true that the subject had been put forward at various times and had been well-aired on a number of occasions. When Darwin did refer to the ideas of such earlier writers as his grandfather, Erasmus Darwin, and Lamark, he was very dismissive and said he obtained nothing from them.

However, Darwin was criticised for making such a claim and he was forced to add to the third edition of his book in 1861 a "Historical Sketch". In this he set out the many papers which had preceded his book on the subject of the transformation of one species into another.

THE EARLY GREEK THEORIES

Most writers agree that virtually all of Darwin's ideas had been proposed very many years before he set sail on the *Beagle* and refer to the works of Buffon, Lamarck and others. What is sometimes not fully appreciated is that as far back as the Greek philosophers, various strange ideas were held on how men came from animals.

THALES (c 600 B.C.) was the first known Greek philosopher who studied the realm of natural history. He believed that water was the cause of all things, and that the earth floated like a vast disc with water around all sides. It is interesting that Nordenskiold, in his book *The History of Biology*, notes the similarity this has with the account in Genesis that the firmament divided "the waters which were under the firmament from the waters which were above the firmament", and comments "That we are here dealing with a theory of oriental [!] origin seems beyond all doubt."[29p11]

ANAXIMANDER (c 570 B.C.) was possibly one of his pupils, who taught that everything had come from "apeiron", the precise meaning of which is uncertain but probably meant some non-material psychic realm. From this arose heat and cold which gave rise to water, which in turn produced earth, air and fire. The Earth condensed out of water to a mud, and from this arose animals and plants. Human beings later arose in the sea,but subsequently cast off their fish skins to walk on dry land. His theories are clearly the very first outline of a primitive form of evolution.

He even claimed that the whole universe returned to the original "apeiron", which then gave rise to another universe in an endless cycle - a conception not unlike the theories of "Continuous Creation" proposed by the astronomer Professor Fred Hoyle only a few years ago.

DEMOCRITUS (c 400 B.C.) may be regarded as the first thoroughgoing materialist, for he taught that there is nothing but atoms and space, and that everything happens through cause and effect.

ARISTOTLE (384-322 B.C.)wrote numerous works which dealt with the whole of the world of nature as he knew it. He made very detailed investigations of numerous forms of animal life,and provided a comprehensive scheme of classification, claiming all the various forms were linked in an evolutionary process. So consistent, complete and dogmatic were his ideas that they dominated all intellectual thinking in the fields of both science and philosophy until the Middle ages. It was Galileo, who by showing that Aristotle's ideas were not in accord with the results of practical experiments, was to be the first in the field of the scientific investigation of natural phenomena.

SPONTANEOUS GENERATION

The well attested 'fact' that rotting meat bred maggots and flies and that various insects arose from mud and slime was sufficient to convince people of the ancient world that decaying or inert material was able to generate life spontaneously. This idea was accepted as a fact until comparatively recent times. As the Christian faith became more accepted, many believers looked upon the idea of spontaneous generation with a good deal of scepticism, although they had no evidence to contradict it. Both Gregory and Augustine spoke out against it as untheological, being refuted by God's word. It was not until Louis Pasteur made his famous experiments proving that the maggots resulted from eggs laid by flies, that the whole idea was finally discredited.

It is sometimes thought that this false concept was only believed by those who were ignorant of elementary facts of nature. Yet one of the most vigorous opponents of Pasteur's conclusions was non other than Professor Haeckel of Jena University, a fanatical propogandist in Germany for Darwin's new theory of evolution. The reason for this attack is very illuminating. He did not criticise the experimental evidence, but only the fact that if accepted, it left him with no option but to believe in a Creator! In England, the ardent rationalist, H.C. Bastian (1837-1915) spent his life similarly trying to discredit Pasteur.

Indeed, I do not think that it is unfair to claim that the modern equivalent of this ancient belief is the well know proposition that life arose from the original "primaeval soup" of the oceans, by chance combinations of atoms over millions of years. It can be clearly demonstrated by simple mathematical probability, that even with the vast periods of time which evolutionists claim, it is still insufficient for even a small protenoid (say 100 amino acids long) to have come into existence. Despite this, it is still seriously propounded in textbooks as having taken place, for the simple reason that the dogma of evolutionary theory demands that it "must" have occurred.

Thus in one form or another, men have wanted to cling to some theory which did not require the activity of or even the concept of God.

CHAPTER 2

THE DARK AGES

It does not take much perception to realise that the desire to evade the authority of his Creator is innate to mankind. This attitude is generally described as being 'scientific', 'impartial' or some such apparently rational approach. This whole subject is so important in understanding why evolution is so readily accepted today that I feel it warrants a digression to examine the way in which it forms a backdrop to the eventual re-emergence of evolutionary ideas at the end of the Eighteenth Century.

EUROPEAN PHILOSOPHY vs. CHRISTIANITY

Following the period of the Greek philosophers, Christianity gradually spread and predominated over many cultures.

It has been well documented by several writers that true science i.e. the discovery of the laws governing the universe, could only arise in such a culture which acknowledged that the universe was created and sustained by a God who was ultimately rational and all-powerful, and who ensured that the laws which nature obeyed were constantly in operation. To the pagan, events occurred at the whim of one or more capricious gods, who needed to be placated by sacrifices, gifts and obeisance. To the Christian, however, the world could now be explored, for his confidence was in a God who had established a universe which worked according to constant laws. The rare occasions when God did temporarily "amend" the natural laws (called miracles by the witnessing mankind) did not alter in any way the *general* working of these laws.

What is important to remember is that most Christians, right up to Darwin's time, fully accepted that these laws (and all the various forms of life) had been called into existence by God, just as they were described in the first chapter of the record he has given to mankind of these events

The rise of the scientific era, however, was delayed for many centuries, for the simple reason that the spread of true Christianity was quickly succeeded by a form of Christianity which was based on a faulty theology, and which adopted many of the pagan customs of the day. For example, in England, Christianity first came to England with the Roman legions, and in A.D. 563 the community of Iona was established. Its evangelical zeal lasted for many generations, but a different form of christian faith entered the country and competed for the approval of King Oswiu at the Council of Whitby held in A.D. 664. After that date, England fell under the sway of ecclesiastical superstition from which she was not freed for many centuries.

In a somewhat parallel fashion, evangelical Christianity continued to hold out in many of the Swiss canton fastnesses, where the flame was kept alight despite bitter persecutions. It was not until such men as Luther in Germany, Bilney in England, and others, were to rediscover the authority and reliability of the Bible that the renewing of the true Christian basis of faith arose which was called the Reformation.

As one traces back the main crises which overtake men from time to time, both as individuals and as nations, the cause can invariably be traced to an inadequate view of life, or a serious error of judgment which may have preceded the crisis by many years. Indeed, the more fundamental the error, the more disastrous is the calamity which eventually overtakes them.

Thus it is with the inadequate theology, which in various degrees plagues the present-day Church, thereby ultimately affecting the way of life of many nations.

The root of many of the problems which we see today has been most penetratingly analysed by Francis Schaeffer in several of his works, most notably in his *Escape from Reason* [40]. In this work, he points to the teaching of Thomas Aquinus (1225-1274) as opening the door to a line of thinking which was to result eventually in today's godless secular philosophy. Aquinus contended that the world of "nature", which in his day was almost completely ignored, should be given much greater emphasis. In so doing, however, he invested it with a certain degree of autonomy, making the study of it independent of its creator. Schaeffer shows how this "freedom" of nature grew to such an extent that first of all it was independent of God, then in opposition, until it finally destroyed the concept of God. He traces the line of thinking from Aquinus through Kant and Rousseau to Hegel. It was Hegel who completely altered the whole foundation of philosophical thought away from basically rational logic to that of his thesis - antithesis - synthesis concept. This philosophy eventually gave rise to the modern day culture which, with its expression in forms which become increasingly grotesque, can be seen to be ultimately without meaning.

This concentration upon the role of nature to the exclusion of God, however, presented the natural philosophers with a serious problem. Having eliminated God, they were left with only atoms and forces, and therefore man, who in their view consisted only of atoms, was ultimately of no significance. Yet they felt that man *was* significant. In order to combine these two concepts, Kirkegaarde (1813-1855) proposed that man "authenticated" his existence simply by taking action. It was his views which gave rise to the "existentialist" school of thought.

Another problem which secular philosophy tried to solve was that if there were only atoms, how could there be any "universal law", which could relate the many aspects of life into one "conceptual whole"? What could be considered as an attempt to solve this was to claim that evolution provided a satisfactory "scientific" explanation of the rise of life and then to invest it with "meaning" and "purpose". It was thus elevated to a "faith" and was used to interpret developments in many spheres of life. In our schools in England, for example, under the heading of "Combined Studies", the subjects of religion, biology, history, social sciences, etc., are all taught as having "evolved" from "simple" beginnings.

Thus we see that, as well as Biblical Christianity, there has always been an alternative philosophy of life towards which men will by nature gravitate. Such views, however, did not achieve any degree of pre-eminence until the Eighteenth Century. It was in the writings of Rousseau and Voltaire, that these ideas were most clearly expressed, the practical outcome of which was to culminate in the French Revolution and in the rebirth of the idea that man evolved from a primeval nature, independent of any need for a God.

CHAPTER 3

THE EIGHTEENTH CENTURY

As European culture emerged from the Dark Ages, men began to study the world around them. In this section we will briefly consider those whose work was to play a part in the eventual rise of evolution.

LINNAEUS (1707-78) was the renowned classifier of plants and animals. He was a creationist, and began to collect specimens in Sweden to determine what were the originally created 'kinds' of Genesis (or species), and what were only varieties. To each of his species he gave a specific Latin name. As his collection from other countries grew, he realized that the most probable unit of creation was the genus. He therefore gave the name of each species a preceding Latin name designating its genus. This system is still in use, although what he classed as an animal species would today be considered a family. His genus in plants still remains.

What is important to remember is that Linnaeus classified the living species *purely for identification purposes*. He never intended that his groupings should be interpreted as providing evidence of evolution between them. In spite of this, his classifications have been used to support the evolutionist's case. In a similar way, what he classed as separate species were sometimes found to be only varieties. The fact that one 'species' could be shown to develop into another 'species' was then hailed as evidence for evolution instead of the classification being corrected. What is certain is that there is no evidence of evolution between the major groups of orders or classes in the animal kingdom, and this has proved a major obstacle to the theory.

BUFFON (1707-88) may be regarded as the first person who had made a scientific study of nature in order to set down a general theory of evolution. As a member of the Academie Francaise, he met many of the philosophers of his day, and his major work was a publication consisting of forty four volumes. In these he described and classified almost the whole of the natural world and provided an evolutionary explanation of their origin. He was the first to claim that geological strata were a result of definite stages in history, and that the planets were the result of a collision between the Sun and a comet. His idea that certain species had been lost allowed the study of palaeontology to proceed with a ready-made explanation of the large gaps between the different species whether living or fossilized.

Fig.1. Lamarck

Fig.2. Erasmus Darwin in 1770

Fig.3. Robert Malthus

Fig.4. Adam Sedgwick

LAMARCK (1744-1829) studied and wrote about the identi-fication of plants. He was appointed as tutor to Buffon's son, and in this capacity visited many botanical gardens in Europe. Although he had been formerly a specialist in botany, he was appointed after the French Revolution as Professor of Zoology in the newly-formed French equivalent to the Natural History Museum.

By the end of the 18th Century, the study of chemistry and physics showed that valuable results could be obtained by careful and detailed experimentation. Lamarck on the other hand wished to propound broad all-inclusive speculations in which man would have an allotted place within his proposed scheme. He therefore feared that this new development might result in the breaking up of scientific study into isolated sections, with no coherent system between them. He therefore published his evolutionary theories which covered many fields of science. This arose from a desire to discover a universal Law which would incorporate all the various sciences into one great systematic materialistic framework - a subject to which I have already referred.

Lamarck's view of evolution involved the idea that there were "subtle and ever-moving fluids" which were "excited", and by which the organs of animals were gradually adapted to their environment. These acquired characteristics were then passed on to their offspring.

He considered that the continents had been built up by inundations of the sea and, even more important, that Nature had had an unlimited amount of time in which to accomplish her tasks of transforming living organisms.

Lamarck's proposal that "fluids" within the animal had sub-consciously modified its organs to adapt to the new habits and environment is essentially a *biological* explanation of evolution, for it deals with the animal as a modifiable organism which has instincts and habits. Today, such an explanation appears non-materialistic. But Lamarck and his contemporaries really believed that these "fluids" existed within organisms. He was therefore *in his day* a thorough-going materialist, that is to say he provided a materialistic explanation for the present forms of life. However his theory is opposed to the current view of evolution called "Neo-Darwinism". This claims that the environment *selects* from numerous *chance* mutations those which are most likely to enable the animal to survive. Lamarck's theory was therefore badly discredited. It does however still have some interesting reper-cussions even today.

It is a fundamental principle of modern evolutionary thought that all explanations must ultimately be materialistic, that is to say all explanations must be based upon accepted laws of Chemistry and Physics. Lamarck's *biological* theory cannot be explained in accordance with such laws and it is for this reason that many evolutionists are fundamentally opposed to Lamarck's views. Indeed, amongst Anglo-Saxon workers in this field it is standard practice to discredit views other than Neo-Darwinism simply by labelling them "Lamarckian".

In his book, Nordenskiold is noticeably generous in his treatment of those who furthered the idea of evolution, for although he often admits that they indulged in wild speculations, which were later found to be wrong, he invariably says that they "provided fresh ideas" or "stimulated research". In his review of Lamarck's strange career, however, he says he was "a discharged lieutenant without any scientific grounding, who from being a Bohemian literary hack works himself up to lasting fame as a scientist and who at the age of fifty becomes professor in a subject that he had never studied before" [29p318].

GEORGES CUVIER (1769-1832) was a protestant of Hugenot descent. Such was his genius that despite the anti-religious views of the nation's leaders after the French revolution, he was appointed to the position of Professor of Natural History at the College de France. He made a diligent study of living and fossil animals, such that he was able to name an animal simply by examining one or two bones.

He flatly rejected Lamarck's ideas on evolution, claiming that the fossil evidence did not show a consistent slow transformation of one species into another. His view was that species were stable but that all living creatures were completely destroyed by various catastrophies, the animals then being replaced by newly created species. He was convinced of this as the strata showed signs of violent upheavals, species disappeared, many never to be replaced again. Furthemore even in his day he was aware that many nations possessed a legend of a mighty flood which destroyed all living animals.

ERASMUS DARWIN (1731-1802), the grandfather of Charles Darwin, was amongst the first in England to set out the broad outlines of the theory of evolution. Indeed, it is remarkable how similar they are to present-day views. His ideas are contained in two works - *Zoonomia* and *The Temple of Nature*. In these he refers to -

a) the formation of the universe from an initial explosion, producing millions of suns and their planets (i.e. other earths?). The modern 'scientific' version of this is the "Big Bang" theory.

b) the 'birth' of the moon from the earth, where the 'South Sea' is now. A mathematical basis for this view was suggested three generations later by Sir George Darwin, Charles' son.

c) the transformation of animals from one species to another, brought about by 'the three great objects of desire' -
   - Sexual drive and selection.
   - Hunger (the struggle for survival.)
   - Security (or the means of escaping or evading capture.)
d) the continual progress of such transformation by natural means without any necessary direction by God.

The similarity of his evolutionary ideas to those of the present day is obvious. He did not, however, produce any consistent evidence to support his views as his grandson later sought to do. He frequently expressed his views in poetry of a particularly forceful and imaginative form, which gained him an influential reputation as a poet, and no doubt greatly assisted the propogation and absorption of evolutionary concepts by his unwitting readers.

ROBERT MALTHUS (1766-1834) was born into a prosperous family. His father was particularly interested in philosophy and agreed with writers such as Rousseau who foresaw the perfectability of mankind. Malthus disagreed with this particular aspect and in 1798 published (anonymously) his famous work entitled *An Essay on the Principles of Population as it affects the Future Improvement of Society.*

Much emphasis is laid upon this work in providing Darwin with the vital concept that "favourable variations would tend to be preserved and unfavourable ones to be destroyed". This whole concept of "survival of the fittest" in one form or another had already been used to support the evolution theory by both Erasmus Darwin and Lamarck. Darwin had read both these writers but claimed he had got nothing from them in forming his theory, yet they clearly referred to 'competition' in the world of nature affecting the species.

What is perhaps not appreciated is that Malthus specifically wrote *against* the idea that the human race would progress naturally to a utopian situation, and the main points of his essay were:-

a) Plants and animals can reproduce and expand in an 'arithmetic' progression. (An arithmetic progression is one in which a constant is *added*, e.g. 1,3,5,7,9,11...)

b) Human beings can expand in a 'geometric' progression - which is faster than 'arithmetic'. (A geometric progression is where a constant is a *multiplier*, e.g. 1,2,4,8,16...)

    c) Therefore the human population would outstrip the available food resources for which they would then have to compete.

    d) It had been clearly shown that *there were limits to which any population could be varied by breeding*.

    e) From this, the human race would *not* progress towards a perfect state but would be kept in check by "misery, vice and moral restraint".

The nearest that Malthus comes to 'survival of the fittest' is a reference to primitive tribes such as the North American Indians whose life is so hard that the weakest members are unlikely to attain manhood.

    Darwin, ignoring the fact that there *is* a limit on varieties, turned Malthus' argument on its head and claimed that in the struggle, the fittest would survive to produce a continual progression of improving species! It would seem from this that Darwin actually used Malthus' idea in a sense opposed to the meaning which Malthus himself intended.

    In order to support his views, Malthus, in a second edition, collected various data to show mathematically the available food supply at the end of one, two and three centuries. His mathematics, however, were quite spurious, for it is obvious that human beings do *not* reproduce as rapidly as animals and plants normally do. Indeed, one single plant can produce many thousands of seeds within its lifetime, whilst the offspring of humans is usually in single figures. Thus, on this basis alone, Malthus' simple analysis is completely refuted. Malthus' main contention is that the area of cultivation will only grow in an arithmetic fashion, and, in this country at least, the population it could support would be outstripped in less than fifty years [24p6]. However, to arrive at this conclusion regarding the "strong law of necessity", he makes assumption upon assumption, with virtually no justification by well documented facts. His figures are little more than pure guesswork, and his conclusions, which form the whole basis of this large work, are thereby utterly invalid.

    It is admitted in the biography appearing in the Encyclopaedia Brittanica that even in his massive sixth edition in 1836, he never "adequately set out his premises or examine their logical status. Nor did he handle his statistical material with much critical or statistical rigour..."[17].

Had Malthus been a clergyman (as he was) who was unskilled in mathematics, the fact that he made some grossly inaccurate predictions, whilst being unsatisfactory to say the least, would nevertheless have been understandable. Such an exoneration however cannot be applied to Malthus, for he was educated at

Cambridge where he distinguished himself in Classics *and Mathematics!* That he should have made some false assumptions is bad enough. That he should furthermore have used fallacious mathematical arguments to 'prove' his theory despite his training is surely quite deplorable and falls far short of the high principles which one might expect from a church leader.

In reading his work, one is struck by the very anecdotal evidence he uses, for he quotes the remarks of captains, sailors, missionaries, etc., about the low state of savage tribes around the world. These conditions may well have been for reasons other than those which Malthus assumed. Furthermore, in the introduction to his second edition, in which he provided some population figures to support his contention, he makes the ludicrous statement —

"I have taken as much pains as I could to avoid any errors in the facts and calculations which have been produced in the course of the work. Should any of them nevertheless *turn out to be false*, the reader will see that *they will not materially affect* the general scope of the reasoning." [24]

He thus puts his speculative theory above any facts which may contradict it!

Ridiculous as his mathematics were, Darwin and his colleagues nevertheless accepted them as being fully reliable, revealing a woeful lack of mathematical ability. These aspects of anecdotal evidence, absence of mathematical rigour and circumventing awkward facts, so exactly mirrors Darwin's approach that we will examine them when we consider his "Origins".

Darwin's account of his reaction to reading Malthus is recorded in his autobiography:

"In October 1838...I happened to read for amusement 'Malthus on Population'... it at once struck me that under these circumstances favourable variations would be preserved, and unfavourable ones destroyed....Here then I had at last got a theory by which to work; but I was so anxious to avoid prejudice, that I determined not for some time to write even the briefest sketch of it."[2p83]

Darwin gives the impression that his reading of Malthus' paper was almost an accident but the result was like a great light dawning upon him. This claim however has been shown to be false, for he had just previously read an article in which Malthus' ideas had been clearly set out. He was therefore fully aware of what Malthus was saying before he read his paper. Indeed, investigators have revealed:

"...the misleading impression Darwin created of the genealogy of his theory. It too was a mask. Darwin almost certainly read Malthus not 'for amusement'... but in the course of persuing the subjects for which his M and

N notebooks had been opened."[60p191]

Even his note recording his reading of the paper is not marked with any emphasis as he often did with other ideas which struck him.

Why should Darwin falsely claim that Malthus' paper was such a turning point in his formulation of the theory of evolution? One cannot be certain, but I would suggest that it performed two functions. Firstly, it freed him from any debt to the ideas of other evolutionist writers who had proposed "survival of the fittest" in one form or another. Secondly, it was a well known paper which was accepted as proven fact, particularly in view of the 'incontrovertible' mathematical basis on which it purported to rest. Such credentials would greatly assist Darwin in his search for evidence which could give credibility to his very speculative theory.

Malthus envisaged such a struggle for food that mankind would only survive in a bare state of existence. His ideas, however, were quite wrong, as is now acknowledged, for such conditions have clearly not come into being. Darwin, living in a nation of rising affluence (thus flatly contradicting Malthus), was completely blind to this. Writing from the comfort of his large country house, he lamented to Lyell the way in which Malthus' 'clear arguments' had been treated for he said:

> "By the way what a discouraging example Malthus
> is, to show during what long years the plainest case may
> be misrepresented and misunderstood." [6p317]

Unfortunately for the poor of this nation, Malthus' essay was not just an incorrect theory. Malthus was made a Professor of History and Political Economy, and the impact of his ideas upon the administration of the country was considerable, being incorporated into the current theoretical systems of economics.

He recommended that the poor laws, with their dole, should be abolished, as they encouraged large families, whilst life in work houses should be 'hard', so that they would be refuges only for those in the severest distress. As a result, many suffered appallingly, whilst traditional charitable giving was discouraged [17].

This essay was used by Darwin to support his theory of evolution, a theory which is fundamental to the Marxist cause. Yet it was this same essay which played its part in bringing about the very conditions amongst the poor which Marx was to use in flaying the (Christian) Capitalist system, i.e. the application of "survival of the fittest" generated the living conditions which were condemned by the very people for whom this was a dogma in their philosophy! It may well be that Marx dared not delve too deeply into this aspect for he was probably aware that he would have cut off the branch which supported both him and his political theory!

ROBERT CHAMBERS' "VESTIGES"

In 1844, Robert Chambers, journalist and editor of *Chambers Journal*, published anonymously a book entitled *Vestiges of the Natural History of Creation* in which he set out his evolutionary views. In stating his case, however, he sought to placate religious opposition by saying that the development was in the control of "the one Eternal and Unchangeable". Today this view would be referred to as "Theistic Evolution", that is "evolution under God's control". The book was very heavily criticized by many scientists mainly due to the very poor scientific facts the author used to support his contentions. Both Sedgwick and Huxley wrote strongly against it. Darwin, however, whilst deploring the evidence given, nevertheless admired the aims of the author and the various subjects he had brought together in dealing with the theory. He was particularly conscious of the fact that it would help to cushion the shock of his book for in later editions of his *Origins* he said -

"In my opinion it has done excellent service in this country, in calling attention to the subject, in removing prejudice, and in thus preparing the ground for analogous views".

Despite the considerable opposition of most of the scientists of the day, the work was very popular. Four editions were published in the first seven months, and further editions appeared each year for the next ten years. Thus did the uncritical public seize upon the 'new idea' which this book proposed, making them even more receptive to a more 'scientific' treatise of the subject which Darwin was to produce later.

SECTION II

GEOLOGY PREPARES THE WAY

CHAPTER 4

PROVIDING THE GEOLOGICAL TIME SCALE

The time scale provided by the Biblical account of history of a few thousand years was far too short a period for the alternative theory of evolution to have operated, for this required time measured in terms of millions of years. Time of this order was provided by the Uniformitarian Theory of the earth's development, the chief evidence to support it being provided by Charles Lyell.

The Uniformitarian theory proposes that the various sedimentary strata were laid down slowly over long periods of time by the ordinary forces of rain, frost, water, etc. acting as they do today. The surface of exposed strata are broken up by weathering, the debris being swept into lakes or the sea, eventually settling out to form the various strata. Under great pressure and heat, these deposits may change their nature and can become very hard. They are then slowly uplifted to become new land masses, which are in turn worn down by weathering to form new strata deposited in the estuaries of rivers once again. It will be obvious that this continuous cycle of events would certainly take an immense period of time to deposit all the various strata; time which is measured in millions of years on geological charts.

The alternative view is the Catastrophy theory which holds that the strata were laid down under violent conditions of earthquakes, volcanic activity and floods. Some consider that these were local forces, but others contend that all the various strata were laid down during one world-wide flood. It was the Catastrophist theory which held sway until Charles Lyell published the first volume of his book *Principles of Geology* in 1830, in which he provided evidence to support the Uniformitarian theory. Following this, the theory was accepted in a space of only some twenty years, and is still today the standard theory which is taught on Geological courses. There is however much evidence against the theory, and some experts have expressed their dissatisfaction with it. We will first of all, however, trace the history of the way in which the Uniformitarian theory supplanted the Catastrophy theory, for this makes a most intriguing

Fig.5. Charles Lyell as a young man

Fig.6. Sir Roderick Murchison

Fig.7. Herbert Spencer

Fig.8. T.H. Huxley in 1885

story.

JAMES HUTTON (1726-97) was the original proposer of the Uniformitarian theory. Strangely, however, this was only a secondary thesis. His main theory was known as "Vulcanist" which emphasized the importance of heat on the earth's formation. This was opposed to the "Neptunist" theory which emphasized the mechanical and chemical action of water. It should be noted that both these theories were catastrophic in basis. Hutton's alternative Uniformitarian theory was largely ignored until Lyell published his book in 1830.

CHARLES LYELL (1797-1875) qualified as a lawyer, but being of independent means, did not practise after 1827. He became interested in Geology whilst at Oxford, and this remained his chief activity for the rest of his life, making tours of Europe and America in its pursuit. He joined the Geological Society of London, was made secretary of the Linnean Society when 26 years old, a Fellow of the Royal Society when 29 and published his *Principles* when 33.

His meteoric rise to acceptance would at first sight seem to suggest that he possessed considerable expertise in Geology. However, the real reason for this seems to have been that Geology was, at that time, in "so feeble a state, that it could not resist the determined assault" of Lyell's theory [1 p71].

The weakness of the field of geology and the corruption of the universities is illustrated by the manner in which holders of professorial chairs obtained their position.

ADAM SEDGWICK was elected Professor of Geology at Cambridge in 1818. His main opponent was Goreham who had studied Geology. However, as he came from a small college, known for its evangelicalism, and had few friends, Sedgwick's intrigues easily ensured that he was outvoted. When elected, Sedgwick boasted, "I had but one rival, Goreham of Queens', and he had not the slightest chance against me, for I knew absolutely nothing of Geology, whereas he knew a good deal - but it was all wrong!" How Sedgwick could know from his ignorance that Goreham was "wrong" is not clear. Sedgwick did however begin to study the subject in earnest, and by 1831 he was made a Fellow of the Royal Society and President of the Geological Society. It was in this year that Darwin first met him and became interested in the subject of Geology.

SIR RODERICK MURCHISON (1792-1871) similarly took up

Geology as a hobby. He resigned from a commission in the Army and spent much of his time fox-hunting and socialising. In an attempt to win him away from a somewhat dissolute life, his wife persuaded him to study Geology, and he attended one of Lyell's lectures. He became very active in the field, being made secretary of the Geological Society of London and eventually President of the Royal Society.

He was a man of enormous energy and in 1820 he was chosen by Lyell to accompany him on a tour of Europe. Many years later he visited Wales with Sedgwick. His main work was the definition of the Silurian, Devonian and Carboniferous strata. These are important divisions of the geological strata which are still used today. One would therefore presume that Murchison, in laying the foundations which were to define these strata would have meticulously checked and confirmed his evidence in the course of his travels. Yet it appears that his approach was quite the opposite. Although his biographers contend that his examination of the Geology of Wales was more painstaking than Sedgwick's [10p103], Darwin wrote in 1843 -

"Murchison and Count Keyserling *rushed* [emphasis his] through North Wales the same Autumn and could see nothing....I cross examined Murchison a little, and evidently saw he had looked carefully at nothing" [2p332].

Similarly it is recorded that in his tour of Russia, he went "at a gallop and geologised while his carriage crossed the Urals behind horses sweating under the lash" [10p109], and that "A fair amount of his work, however, later proved to have been somewhat superficial and unreliable".[12p173]

Statements such as these give one little confidence in geological theories based upon such suspect evidence.

Interestingly, towards the end of his life, Murchison reverted to a belief in convulsions, opposed the views he had previously worked out with Lyell, and considered there was no evidence which supported Darwin's theory.

CHAPTER 5

DARWIN'S GEOLOGICAL THEORIES

Darwin first became interested in geology in 1831 during his last year at Cambridge, where he became friendly with Sedgwick. He had so little interest in the subject however that he never attended any of his lectures. He did not study the subject in depth and his only field experience was a visit to North Wales with Sedgwick.

Darwin, under the influence of Sedgwick, was at first a convinced catastrophist. Just before he sailed on the *Beagle* however, his friend, the Rev. John Henslow gave him a copy of the first volume of Lyell's *Principles of Geology* with the advice that he should not believe its contents! As he read the *Principles*, he seized upon the theory with all the zeal of an enthusiastic amateur and arranged to have the following two volumes sent on to him as they appeared. His very inadequate training in geology would have made it impossible for him to assess the validity of the views which Lyell propogated. Nevertheless, Darwin, based all his interpretations of the geological formations which he examined during his voyage on the Uniformitarian theory. So confident was he of his knowledge of Geology that no sooner had he examined the rocks on his very first port of call, the Cape Verde Islands, than he considered writing a book on the subject of the geology of the countries he would visit.

Darwin's rapid acceptance of Lyell's theories is particularly strange as he commented in his diary that "everything betrays marks of extreme violence". Similarly, having experienced an earthquake in Chile, he was particularly impressed by the enormous amount of change that could take place in a short period of time. Nevertheless, after his return to England, when describing the earthquake in his official *Journal* of the voyage, he played down these violent effects, and described it as "a paroxysmal movement, in a series of lesser and even insensible steps". This was not an isolated instance of Darwin's published works being wildly different from his accounts of what actually took place, as revealed by his diaries and letters *written at the time*. This is well documented by Himmelfarb, and its importance in the history of Darwin's evolution theory will be considered later.

It is far from clear why Darwin should have accepted Lyell's views so enthusiastically. However, it would seem probable that what Darwin found so attractive in Lyell's writings was a characteristic which he himself possessed to an inordinate degree, namely the tendency to make wide-ranging generalizations and theories based upon little evidence. Regarding Lyell's abilities, one minor factor which may have had some influence was his exceptionally poor eyesight, which became worse with age, to such an extent that he had difficulty in examining the data and drawing maps, some of this work being done by assistants.

Coral reefs

Darwin's own habit of theorizing is well illustrated in his ideas of how coral-reefs had developed. Even before he had read Lyell or seen a coral atoll, he had deduced that they must have been formed around land or mountains which were slowly subsiding. Himmelfarb mentions that some time after he had published a paper on his theory in 1842, it was proven that in fact they also form around land that was actually *rising*, thus demolishing all his plausible and carefully marshalled arguments. Darwin nevertheless stubbornly clung to his theory. Today, geologist consider that Darwin's theory may be feasable but they are far from certain of the precise cause of the formation of coral reefs.

Another example of Darwin's inadequacy as a geologist concerned his views about certain formations in Glen Roy in Scotland which he published in 1839. He subsequently admitted however that his paper was "one long gigantic blunder from beginning to end".

I am not criticizing Darwin merely for propounding a theory which was later disproved, for this is perfectly admissible. What should be emphasized, however, is that this shows a tendency to formulate ideas on a grand scale based upon a minimum of evidence. With regard to his coral-reef theory, Darwin had formulated his ideas on the basis only of known facts about coral-growth. All this had been done before he had seen a single coral reef for himself! Thus, all his theorizing about sinking land could have been worked out in the comforts of England, for he admitted he had no *direct* evidence to support his ideas.

Darwin indeed was more interested in the rise and fall of land-masses than in uniformitarianism, which is rarely mentioned in his diaries. It was only after this voyage, when he needed the vast periods of time Lyell provided, that he used the theory more fully.

CHAPTER 6

FORTUNES OF THE UNIFORMITARIAN THEORY

THE THEORY ACCEPTED

We have shown in the previous chapter the way in which the case supporting the Uniformitarian theory was presented by a small group of amateur gentlemen of independent means. That Lyell had produced three large volumes supporting the theory certainly suggested that the case was based upon thoroughly researched evidence. However, as will become clear in due course, the case actually rested upon the imposition on the fossil record of a preconceived idea, any contrary evidence being deliberately ignored or made light of.

It must therefore be asked how this theory called "Uniformitarianism" came to predominate when all the leading geologists were convinced catastrophists. Three factors combined to bring this about.

Firstly, it has already been noted that the very amateurish 'science' of geology was in no state to rebut by means of well-documented evidence the claims of a system which had been skilfully argued and propounded with the extremely confident approach which Lyell and his colleagues adopted.

Secondly, in the 1830's geology became a very fashionable pursuit, followed by many thousands of amateurs. It was a healthy outdoor activity, no expensive equipment was required, and the basic principles were easily grasped. Lyell lectured to audiences of thousands of people who were keen to examine the strata in the light of the new theory.

Thirdly, Lyell, being an expert lawyer, knew how to present his case in such a skilful manner, as to lead his readers into accepting his theory without noticing its weaknesses and inconsistencies. Furthermore, as he had inherited a large fortune from his father, he was able to tour Europe and America, collecting geological evidence for his theory. This made it seem extremely impressive. With the very poor salary earned at that time by those in high positions in the field of geology, few who opposed him would have been able to afford to collect an equally impressive array of samples as evidence in order to contradict his theory. One or two who did, however caused Lyell considerable difficulty in trying to overcome their objections.

With this combination of circumstances, Uniformitarianism rapidly became accepted both by the interested public and by the teaching establishments. It is indeed an example of a theory owing its popular acceptance not to the facts supporting it, but rather to the way in which the intelligentsia, who, lacking any specialist knowledge, could be induced to accept a particular interpretation of the carefully selected facts presented.

THE THEORY ATTACKED

The most important question which must be considered at this point is whether Lyell's theories were correct. It is not intended in this work to give a detailed scientific refutation of the Uniformitarian theory, but it will suffice to show how, even in his own day, there was sufficient evidence available to prove that the theory was very inadequate.

His first volume was reviewed by Whewell, a noted expert of his day who, amongst other things, wrote about the philosophical aspects of scientific theories. In his review he made the scathing comment that Lyell's ideas "must speedily fall back into the abyss of past fantasies and guesses from which he has invoked it in vain."[27 p293]

Elie de Beaumont

Very shortly after the publication of the first volume of his *Principles*, Lyell read the geological theory of the Frenchman Elie de Beaumont which contradicted his own views. Elie de Beaumont claimed that whilst there were long periods of tranquil deposition, mountains were upthrust by violent forces.

In discussing the whole controversy, Wilson, in his biography of Lyell, says that the the "...theory brought together a broad range of facts about the structure of the Alps..." and that it was "...supported by an engaging array of geological evidence, some of it new...". He refers to the vigorous way in which Sedgwick championed the theory in his presidential address to the Geological Society, in the course of which he said that it was "...based on an immoveable mass of evidence..." and that Sedgwick agreed with the conclusions because they were "...not based on a priori reasoning, but on the evidence of facts..."[27 p350]

Wilson, however continues-

> "In attempting to counteract catastrophism Lyell
> was at a fundamental disadvantage because the basic
> appeal of the catastrophic doctrine was not rational but
> emotional. It seemed to offer a new prop to the crumbling
> foundations of religious cosmogony. It was not that
> Sedgwick would admit that there was anything whatever
> wrong with the foundations of religion, but the very

speed with which he accepted Elie de Beaumont's sweeping theories is suggestive of an underlying anxiety. [!] In judging the two doctrines there was no critical piece of evidence to which one could point to decide which was true. Both carried a vast array of geological evidence in their train. Lyell could show that particular pieces of evidence which Elie de Beaumont had used were not sound, or that they could be interpreted in another way. This, however, was only to whittle at the edges of the theory and would do little to dislodge the central faith which, he said, experience had shown to be more productive of new discoveries and new truths in geology. He was keenly aware that the issue was one of opposing faiths". [27p350].

Thus Wilson follows in Lyell's footsteps in decrying the creationist's case by claiming that it is "...not rational but emotional...". He completely contradicts this by later saying that both views could present a vast array of geological evidence! What is even more revealing is his clear admission that the whole subject was a matter of *faith*. Lyell was extremely careful in his wording of his books. Indeed he spent over a whole week simply on the wording of the first chapter of his third volume. He always sought to infer that, unlike that of his opponents, his case rested upon incontrovertible reasoning which was deduced from observed facts. Wilson's honest admission that really the whole matter was a subject of *faith* completely destroys the inference which Lyell had so painstakingly tried to construct.

Edward Charlesworth

A further attack came a few years later. One basis of Lyell's case rested upon the interpretation of the position of the numerous types of sea shells (molluscs) which appear in the strata. Lyell claimed that a stratum bearing a large percentage of molluscs which were the same as species still living, showed that it could have only recently been laid down. Conversely, one with a very low percentage of species still living would be ancient. In France, Gereard Deshayes had worked on the editing of Lamarck's works but had fallen on hard times. He possessed one of the largest collections of molluscs in Europe and was an expert in the subject. Lyell arranged to pay him so that he could collate the shells with the strata in which they were found and thus provide evidence of the increasing age of the strata with depth for the Tertiary period.

In 1835 Edward Charlesworth made a careful study of a crag along the Suffolk coast and read a paper to the Geological Society in which he showed that the crag was older than the age given to it by Lyell, and that furthermore it consisted of two different strata. In addition, he cast doubts upon the whole principle which Lyell had used in relating the date of a strata to the ratio of living to extinct species of molluscs. He suggested that some of the shells had been washed out of one strata and redeposited in the other, thus ruining the basis which Lyell used for his dating method.

Much of the subsequent discussion centered around the identification of which molluscs were extinct and which were still living, with experts from each side giving remarkably different figures. It is not necessary to give all the various arguments for the main interest is the way in which these were conducted and the eventual outcome. One example however will be given to show how widely opinions differed. In a second paper delivered to the British Association in 1836, Charlesworth claimed that whereas Deshayes had identified 40% of the crag shells as species that were still living, another expert had said that *more* than 40% were still living. However, three other experts had said that they were all extinct! With such a wide range of opinion, Charlesworth claimed that Lyell's method was invalid.

Lyell took these attacks very seriously, travelling to the continent to study the collections of other experts, consulting his friend Deshayes, and making more than one tour of the Norfolk and Suffolk coastline.

The outcome of all this was that Charlesworths claim that the crag was two strata with one of them much older than the date given to the formation by Lyell was vindicated, although Wilson maintains that the final figures showed that the percentage method of dating was confirmed. What is interesting however is the description which Wilson gives of the reactions of the personalities involved, his account taking no less than 26 pages on this particular controversy. Charlesworth's first paper (which strangely was printed in the *Philosophical Magazine*) "was a thoughtful paper based on the observation of several years....it was not very favourably received."[27p465]. A reply came from a Mr. Woodward, "but he was wrong in his facts". After much investigation and travelling by Lyell, "He acknowledged the correctness of Charlesworth's opinion ...but his acknowledgement was not perhaps particularly conspicuous or generous"[27p480] although by publishing a later paper on the subject in 1839 in a magazine edited by Charlesworth, he sought to make amends for his previous "frugal acknowledgement" of the latters contribution to the study of the area.

It would appear that subsequent to producing his first paper to the Geological Society, Charlesworth fell upon hard times for in 1838 Darwin wrote to Lyell saying:

"But poor Charlesworth is of an unhappy and discontented disposition. He is, moreover, very much to be pitied. The Zoological Soc. Are going to give up the Ass't Secretary's place & it is feared that he has a disease of the heart, so that altogether he is greatly to be pitied."[27p481].

As a result of his paper, Charlesworth would have been opposed by many of those who had accepted Lyell's view. That he should have subsequently lost his job and suffered a heart complaint, possibly brought on by the stress of the controversy, would have put him in dire circumstances, and Darwin shows his obvious concern. In view of this, although the letter itself was printed in Darwin's *Life and Letters*, one cannot help wonder why this particular passage was omitted.

CHAPTER 7

LYELL REFUTED

Although we will not be providing a scientific criticism of Lyell's theories, a brief outline however must be given for completeness and for the overwhelming importance which it has upon the whole theory of evolution.

As we have seen, Lyell was opposed in his own day and had to admit that his observation and dating were faulty on one strata which could be closely investigated by other geologists. This does cast some doubt on those areas of the continent which could not be easily inspected by rival experts. Furthermore, the whole subject of the identification of distinct species of molluscs is difficult. Dewar [28] has recorded that the Micraster shells (an extinct sea urchin), occasionaly referred to by evolutionists in support of their case, are simply variations of one species and should not be classified as three separate species. He·also mentions that one species of Dog whelk has nineteen varieties which differ widely in size, shape, roughness, etc. Additionaly, three separately named species of crustaceans were found to be one species. One lived in salt water, one in fairly salty water and the third in fresh water. It would appear from all this that molluscs can vary widely under different conditions, making positive identification of extinct species particularly difficult. If this is the case then Lyell's use of them to determine the age of the Tertiary strata is baseless.

An even more damaging attack upon his whole uniformitarian theory, however has more recently come not from the creationists but from present day evolutionists.

In recent years, several writers have revealed serious doubts not only about the evidence which Lyell provided but his motive for formulating his theories and even his integrity. For example Ospovat has studied the reason why Lyell formed his theory of climate. Lyell had proposed that the climate had been warmer in ages past because the land masses were concentrated near the equator. Cooler periods followed as the land moved towards the poles and then slowly returned towards the equator in a cyclic pattern. When fossil evidence of progress was presented to Lyell Ospovat comments:

> "Lyell however was not dismayed. One of the great
> advantages of his one-cycle theory of climate and life
> was that it could not be tested against any sort of
> evidence." and concluded "Here Lyell's preconceptions

led him to construct a theory of the earth out of distinctly fanciful speculations which were, of necessity, based upon no evidence at all."[58]

One is left wondering why Ospovat should have felt that it was *necessary* for Lyell to base his theory on *no evidence at all!*

Lyell denounced

Another writer who has raised scientific objections to Lyell's theory is the well-known evolutionist Professor Stephen Jay Gould, who is Head of the Department of Geology at Harvard University. In an article entitled *Catastrophies and Steady State Earth* in *Natural History*, he wrote:

"Charles Lyell was a lawyer by profession, and his book is one of the most brilliant briefs ever published by an advocate...Lyell relied upon two bits of cunning to establish his uniformitarian views as the only true geology. First, he set up a straw man to demolish.. In fact, the catastrophists were much more empirically minded than Lyell. The geologic record does seem to require catastrophes: rocks are fractured and contorted; whole faunas are wiped out. To circumvent this literal appearance, Lyell imposed his imagination upon the evidence. The geologic record, he argued, is extremely imperfect and we must interpolate into it what we can reasonably infer but cannot see. The catastrophists were the hard-nosed empiricists of their day, not the blinded theological apologists.

"Secondly, Lyell's 'uniformity' is a hodgepodge of claims. One is a methodological statement that must be accepted by any scientist, catastrophist and uniformitarian alike. Other claims are substantive notions that have since been tested and abandoned. Lyell gave them a common name and pulled a consummate fast one: he tried to slip the substantive claim by with an argument that the methodological proposition had to be accepted, lest 'we see the ancient spirit of specualtion revived, and a desire manifested to cut, rather than patiently to untie, the Gordian knot.'"[25]

Such a strongly-worded criticism is tantamount to a denunciation, not only of Lyell's arguments, but also of his deception, and even his integrity. There is a great deal of evidence that the strata were in fact laid down very rapidly and not slowly as Lyell suggested.

Many geologists today admit that there is considerable evidence that most strata were laid down under catastrophic conditions, completely contradicting the Uniformitarian theory to which Lyell devoted the whole of his life. Although admiting this, modern geologists nevertheless consider that Lyell was right in demolishing the Genesis account, for the long periods that Lyell provided are now supported by the evidence of Radiometric dating. Some of the inadequacies of this method of dating are given in Appendix III. If these are accepted as invalidating the results then the occurance of one world-wide catastrophic flood as recorded in Genesis should not be dismissed upon such 'scientific' evidence.

This does raise the question why Lyell, faced as he must have been with evidence of catastrophism, should nevertheless have deliberately constructed a cleverly-worded series of arguments to prove that the strata were laid down slowly, each one taking a vast period of time.

It is generally admitted that Darwin would never have developed 'his' theory of evolution if Lyell had not already argued that the earth was of a much greater age than expected, for this allowed adequate time for evolution to have taken place. In my view, however, it was Lyell's deliberate intention to prepare the ground for evolution by publishing his *Principles of Geology*, leaving the theory itself to be promoted by someone else at a later stage. How Darwin was encouraged to fulfil this role *after* his return from the *Beagle* voyage will be considered later.

SECTION III

THE RISE OF EVOLUTION

CHAPTER 8

DARWIN'S CONTEMPORARIES

With the rise of the Industrial Revolution and the growing number of scientific discoveries and inventions, times were changing rapidly and men were becoming increasingly self-confident in their ability to eventually control the direction of their destiny. It is generally acknowledged that the "time was ripe" for the acceptance of the theory of evolution, promoted as it had been for many years by earlier supporters. Although it fell to Darwin to produce a 'scientific' treatise of the subject, nevertheles it would have had little impact were it not for the vigorous activities of several men in influential positions, who thrust the theory forward by every available means. In the course of this work we will only briefly refer to some of them in passing. A few however merit closer attention and in this chapter we will give some details of two of Darwin's friends. In the next chapter we will make a more extensive examination of the activities of T.H. Huxley.

JOSEPH HOOKER (1817-1911)

Hooker was the botanist from whom Darwin obtained most of his information on plant classifications etc. Hooker succeeded to his father's position as Director of Kew Gardens which, under his guidance, was to become world famous. He was a very close friend of both Darwin and Huxley but does not appear to have had a strong personality as did say Huxley and consequently he appeared far less often in the public eye. Nevertheless, he was a strong supporter of evolution and wielded considerable influence in the promotion of the theory due to his position in various organisations and groups as we shall see later.

HERBERT SPENCER (1820-1903)

Spencer was a prolific writer of the Victorian era and a fanatical evolutionist even before Darwin published the *Origin*. Every sphere of life - economics, biology, social sciences, etc. etc. - he interpreted as developing upon evolutionary principles. In line with his views, he was also a strong advocate of non-interference in the state of the poor, writing -

> "The poverty of the incapable, the distress that comes upon the imprudent, the starvation of the idle, and those shoulderings aside of the weak by the strong, which leave so many 'in shallows and miseries', are the decrees of a large far-seeing benevolence." [lp346]

His works were very effective in bringing the theory of evolution to the attention of the general public, for he wrote in a popular style. However, he would deduce broad generalisations from what were often unsatisfactory examples. When challenged that one of his examples was only true as an exception, he replied, "It would do just as well as an example". So keen was he to present a wholly materialistic picture of the universe that he strongly opposed well established scientific principles. Such conduct appalled the serious scientists of the day, whilst Huxley quipped that "Spencer's idea of a tragedy was a deduction killed by a fact".

A glowing account of his work in the *Encyclopaedia Britannica* assures us that his knowledge was "buttressed by a detailed, scientific examination of a vast range of biological and social phenomena..." Irvine however in his book *Apes, Angels and Victorians* presents a rather different picture:

> "In later years he was so little capable of the docile passivity necessary for getting through a long book that he could not hear more than a paragraph read aloud without launching on a disquisition of supplement or rebuttal. No modern thinker has read so little in order to write so much. He prepared himself for his *Psychology* chiefly by perusing Mansel's *Prologomena Logicae*, and for his *Biology* by going through Carpenter's *Principles of Comparative Physiology*. He produced a treatise on sociology without reading Comte, and a treatise on ethics without apparently reading anybody. Clubs provided Spencer with an excellent substitute for reading. He pumped the authors themselves. Strolling about midday through Kensington Gardens to the Athenaeum, he lunched with one notability, buttonholed a second, played billiards with a third, rifled the periodicals in the library for facts — and was thoroughly crammed for the next morning's composition."[16p183]

Amongst Spencer's many friends were George Eliot, the novelist and Beatrice Webb. Huxley was a particularly close friend, and in 1858 Spencer moved to St. John's Wood to be nearer to him. He often played fives with him and accompanied him on walks on Sunday afternoons. Huxley tried to restrain some of Spencer's wilder scientific speculations, and to this end, he corrected the proofs of several of his works prior to publication.

Interestingly, Darwin, in a passage which was omitted when his autobiography was first published, commented that he did not like Spencer particularly, considered him egotistical and found his theories very unconvincing [26 p108]. Like Darwin, Spencer suffered from sleeplessness and a variety of other nervous and physical ailments and occasionally adopted some unusual practices in an effort to overcome them. Despite these drawbacks, he produced a formidable number of works during his lifetime.

Today, Spencer's writings are read with no little surprise that such sweeping metaphysical theories, with so few facts to support them, should have been so influential in their day. This does however go to show that the reading public were only too willing to agree with a philosophy which was so congenial to the spirit of the time whether well supported or not.

Somewhat strangely however, Spencer, as ardent a materialist and evolutionist as he was, eventually admitted:

> "The Doctrine of Evolution has not furnished guidance to the extent I had hoped. Most of the conclusions, drawn empirically, are such as right feelings, enlightened by cultivated intelligence, have already sufficed to establish" [1 p356],

and was reluctantly forced to the conclusion that:

> "the control exercised over men's conduct by theological beliefs and priestly agency, has been indispensable" [1 p338].

CHAPTER 9

T.H. HUXLEY

Born in 1825, Huxley was an unknown but ambitious biologist in the early stages of his career. He found that progress was blocked by establishment figures, such as Robert Owen, who kept to themselves the prestige of any new discoveries. In addition the salaries of even very eminent scientists were quite low, particularly when compared with the high incomes enjoyed by the clergy at that time. This made Huxley envious, and he expressed his bitterness in later years when, in answer to the question why he was so vitriolic, he replied:

> "My dear young man, you are not old enough to remember when men like Lyell and Murchison were not considered fit to lick the dust off the boots of a curate" - and added - "I should like to get my heel into their mouths and scr---unch it around" [1 p217].

In his early days Huxley had little sympathy with the theory of evolution. He had strongly condemned the scientific inadequacies of *Vestiges* in a review (although he later regretted the forcefullness of his criticism). Similarly, when he first met Darwin in about 1855, he commented upon the sharpness of the division between species, and was puzzled by Darwin's smile when he disagreed.

What will doubtless surprise many, however, is that although he is recognized as being by far the fiercest propagandist for Darwin's evolutionary ideas, earning the title "Darwin's Bulldog", *he never accepted even to his death that the case for evolution had finally been proven.*

What Huxley realized was that Darwin's theory would enable him to achieve a notable reputation as its most virulent propagandist and at the same time allow him to flay the established churchmen, for he was an "inveterate hater of religion" [11 p90]. When he read the *Origins*, he wrote a very flattering letter to Darwin offering his help, saying,

> "And as to the curs which will bark and yelp, you must recollect that some of your friends, at any rate, are endowed with an amount of combativeness which...may stand you in good stead. I am sharpening up my claws and beak in readiness" [13 p254].

Darwin, in view of the fierce storm which his book had aroused, was only too willing to accept such an offer of help. It is interesting to reflect upon this strange relationship, for Darwin, who at one time was intending to enter the church, was prepared to have his theory aggressively promoted by an anti-Christian, whilst Huxley acted as the most vitriolic propagandist for a theory that he never really believed!

Huxley wrote a review of the *Origins* for *The Times*. In his second biography of Huxley Bibby gives an account of this incident saying:

> "*The Times* sent Darwin's book to one of its regular anonymous staff reviewers, to whom Huxley made the generous offer [!] of helping in the labour of writing while claiming none of the credit".[12p39]

In writing to Hooker, who was the only one in on the secret, he said,

> "...I earnestly hope that it may have made some of the educated mob, who derive their ideas from the Times, reflect. And whatever they do, they *shall* [emphasis his] respect Darwin (& be d - d to them)" [12p39].

It is perhaps of passing interest that the last few words in brackets were omitted from Huxley's *Life and Letters* [13p256].

What is perhaps even more interesting is that in his first biography of Huxley, Bibby gave a noticeably different account of how the review came to be written for he said:

> "So soon as *The Origin* appeared Huxley arranged for a review by himself to be printed in the leading daily *as if by a staff writer*"[50p78].

In this second version of the incident, Bibby is clearly protecting Huxley's image by omitting the fact that this important review in the *Times* was not the result of an offer but was deliberately engineered by Huxley himself to appear as if from an unbiased judge.

His biographers often describe his replies to critics as "a crushing victory", but rather more frequently than is fully credible. I rather suspect that his posture of scientific integrity combined with linguistic fireworks was calculated to intimidate anyone who might attack the theory which he knew only too well was based on inadequate foundations.

Huxley's aggresive ambitions were more than fulfilled when he became Darwin's chief propagandist. Clark says:

> "...it was not long before he discovered that loyalty to Darwin paid him well. He was scarcely known in the scientific world until the British Association meeting in 1860, which at once brought his name to the fore" [11p92].

He became a Professor at the School of Mines, and by sheer hard work, became influential in many spheres of science, turning down many tempting offers outside London, for he wrote, "I will *not* leave London - I *will* make myself a name and position..." [12p25](Emphasis his). Such was his ability that the School of

Mines expanded rapidly and eventually became the Imperial College of today.

Opposition to The Church

As well as flaying scientists opposed to evolution, he also criticized Churchmen for their theological writings, having a heated correspondence with Gladstone in *The Times*. His target was often those who were pillars of the Anglican Establishment, and in some cases, he may well have been justified in pricking the bubbles of a few self-righteous clerics. However, in 1890, he campaigned against the "dictatorial manner in which General Booth was administering the Salvation Army"! As an ultra-dogmatist himself, here indeed was the pot calling the kettle black!

With all his attacks against the Established Church, there was one minister in London of whose fame he must have heard, whom he never criticized. I am referring to Charles Haddon Spurgeon, the brilliant preacher who enthralled a congregation of 6,000 between the years 1854 and 1892. As Spurgeon was known for his ability to defend the strongly held reformed doctrine which he preached, did Huxley realize that in taking on such an opponent, he would have more than met his match?

The dark cloud

Huxley was indeed a complex character, being ambitious, hardworking and authoritative to an excessive degree in each. Yet "Throughout his life he seems to have been subject to maniac alternations of furious hard work and lethargic despondency" [16p17]. He suffered from periods of extreme depression with two very bad attacks. One of these occurred when he was thirteen after watching a macabre post mortem. Another attack occurred during his service as an assistant surgeon in HMS *Rattlesnake*.

Except perhaps for a letter that he wrote to Charles Kingsley shortly after Huxley had lost a dearly loved young son, he hardly ever revealed his deepest thoughts. He usually maintained that he was simply pursuing "truth". Similarly, actual insights into his basic character are few. Irvine, however, does record the observations of Beatrice Potter (later Mrs. Sidney Webb). She first met him in 1886, towards the end of his life, and wrote -

> "He is truth loving, his love of truth finding more satisfaction in demolition than in construction....his early life was extremely sad....For Huxley, when not working dreams strange things: carries on lengthy conversations between unknown persons living within his brain. There is a strain of madness in him; melancholy has haunted his whole life. 'I always knew that success was so much dust and ashes. I have never been satisfied with achievement'" [16p233].

Irvine comments -

> "Perhaps, also, Huxley was afraid to face himself. He
> had found in life no satisfying constructive purpose....No
> less in his fastidious pride could he be content with
> himself morally. Some sense of guilt or impurity - hinted
> at in the famous letter to Kingsley - kept him always at
> his treadmill of self-discipline....More and more through
> these years his triumphal progress looks like a flight from
> reality" [16p234].

The letter to Kingsley, to which he referred, contains the poignant
and revealing statement -

> "Kicked into the world a boy without guide or training,
> or with worse than none, I confess to my shame that few
> men have drunk deeper of all kinds of sin than I. Happily
> my course was arrested in time - before I earned absolute
> destruction - and for long years I have been slowly and
> painfully climbing, with many a fall towards better
> things." [13p318].

Huxley's early life is a little obscure, but Bibby shows how very
stressful his home background must have been. His father ended his
days in an asylum, and his mother died at an early age. Of his five
brothers and sisters, one brother suffered "extreme mental anxiety"
whilst another in his later years was "as near mad as a sane man can
be".

The effect which such a background probably had upon the
clever young Thomas can be imagined. Perhaps here was the spring
of his overwhelming drive for notoriety, a goal for which he was
prepared to labour tirelessly. Fame is indeed a spur, but in excess,
as Huxley must have surely found, it can become an unsparing
goad.

Religious perception

What is interesting about Huxley's philosophical progress is
that, although he continued to write fiercely critical letters and
papers mainly against the Anglican establishment, he became
increasingly aware of the appalling results which could flow from
the application of evolutionary principles in everyday life. He
realized that evolution could never provide an explanation for the
moral instinct within man and wrestled with this problem for many
years. Obviously, as an agnostic, he would not acknowledge the
existence of God. He would therefore be quite unable to find a
satisfactory explanation for the existence of his own innate moral

sense, for he had effectively dismissed the Author from the scene.

Huxley's religious thinking was particularly noticeable after 1888, when he wrote a paper entitled *The Struggle for Existence in Human Society*, which he admitted in private letters was directed against Spencer, although Spencer was not mentioned by name. In 1893 he gave the Romanes Lecture in which he condemned those who claimed that evolution had a moral and ethical content. Indeed such were his views at this stage that he considered the lecture an opportunity to discourse upon "Satan, the Prince of this world". He saw all too clearly that Darwinists had a hopelessly inadequate explanation of evil in the world and wrote -

"The doctrines of predestination, of original sin, of the innate depravity of man and the evil fate of the greater part of the race, of the primacy of Satan in this world, of the essential vileness of matter, of a malevolent Demiurgus subordinate to a benevolent Almighty, who has only lately revealed himself, faulty as they are, appear to me to be vastly nearer the truth than the "liberal" popular illusions that babies are all born good, and that the example of a corrupt society is responsible for their failure to remain so; that it is given to everybody to reach the ethical ideal if he will only try; that all partial evil is universal good, and other optimistic figments, such as that which represents "Providence" under the guise of a paternal philanthropist, and bids us believe that everything will come right (according to our notions) at last"[15 p220].

CHAPTER 10

DARWIN'S LIFE

CHARLES DARWIN (1809-1882) was the son of the well-to-do Dr. Robert Darwin, himself the son of the famous Erasmus Darwin. Charles, knowing that one day he would inherit his father's fortune, never studied seriously at the various schools and colleges to which his father sent him, but spent much of his time with the more disreputable young men of the universities, and developed almost an obsession for the sport of shooting wildfowl. Such was the despair of his father at his conduct that he upbraided him at one stage with the words, "You care for nothing but shooting, dogs and rat catching, and you will be a disgrace to yourself and all your family". Darwin had a very high regard for his father, but this rebuke however seems to have had little effect upon him, for he continued with these pursuits throughout his academic career, until he joined the *Beagle*.

When sixteen, Darwin was removed from his school and sent to Edinburgh University in the hope that he would take up a medical career. Darwin, however, had no real interest in medicine and spent much of his time with his more boisterous friends. He spent two years here, and during this time did begin to take an interest in the study of beetles and some other Natural History subjects. However, when Charles eventually told his father he did not wish to practise medicine, he later recalled that "He was very properly vehement against my turning into an idle sporting man, which then seemed my probable destination".

At this time the religious establishment was in such a low state that it was a recognised 'career' for dullards and failures. Robert Darwin, "having joined the Freemasons as a young man, he later became so secretive about his disbelief, that he had his children brought up in a thoroughly orthodox fashion, even going so far as to plan a clerical career for Charles" [1 p9]. Thus, in seeming desperation, it was suggested to Charles that he enter for Holy Orders. Having no great aversion to the idea, he agreed, entering Cambridge University in 1828.

In the course of his theological studies, he read Paley's *Evidences of Christianity*. This famous work shows how the very highly complex organisms and animals in the world give a clear indication that they were created by an infinitely intelligent God who designed each of them for the precise part he wished them to play in the world of Nature. Darwin was particularly impressed with the orderly, logical reasoning used and studied the work so thoroughly that he said he could have written out the arguments with perfect exactness.

In his last year he became more interested in the Natural Sciences. Science in those days was very much a fringe subject at universities compared with Classics and Mathematics. As has been shown, election to professorial chairs of the sciences depended upon how many friends one had, for scientific subjects required no examinations and were treated as respectable hobbies.

Whilst at Cambridge, Darwin became a close friend of the deeply religious Rev. John Henslow, a friendship which lasted the rest of his life, and it was at discussions held at Henslow's house that he became interested in the Natural Sciences. It was through this connection that he met Sedgwick, whom he accompanied on a visit to North Wales. This was Darwin's only geological field trip before his voyage on the *Beagle*.

Darwin's main pursuits, however, were shooting and beetle collecting. He admitted that his years at Cambridge were "worse than wasted" and that he mixed with "dissipated, low minded young men" whose company he admitted he greatly enjoyed. Thus Darwin, in the summer of 1831, left Cambridge and looked forward to being a country clergyman of perfectly orthodox Christian beliefs, but who would have ample time to follow his interests in shooting, beetle collecting and other such respectable occupations for a man of his 'calling'. These mundane plans for the future, however, were to be completely altered by two letters which awaited him on his return from his visit to North Wales.

The first was a request for a naturalist to go with the Admiralty vessel, the *Beagle*, commanded by the popular Captain FitzRoy. This was enclosed with a letter from Henslow who had proposed Darwin as the most suitable person likely to go. Henslow himself had been asked and he had offered the post to his brother-in-law. However, both had family commitments and so Henslow had suggested Darwin.

Darwin appeared to be quite keen to go, but consulted his father, who pointed out a number of objections to this "wild" and "disreputable" scheme. Seemingly without protest, in view of his father's opposition, Darwin had second thoughts and wrote a letter of refusal. The next day he visited Josiah Wedgwood, with whose family the Darwins had always had a close friendship, since the days of Erasmus Darwin. More than one intermarriage had taken place, and Charles himself was later to marry Emma Wedgwood.

On relating to Josiah the offer he had just refused, his uncle pressed him to change his mind, and both wrote and visited Charles' father to obtain his agreement, which was given. Thus with mixed feelings of elation and considerable apprehension, Darwin eventually sailed from England on December 27th 1831.

CHAPTER 11

THE "BEAGLE" VOYAGE

There is little point in describing the route of the *Beagle* for this is well documented in books dealing with Darwin's life. In this section, therefore, we will examine some of the inaccuracies of the popular accounts of Darwin's voyage and his formulation of the theory of evolution, where they are misleadingly in error.

When, in due course, the voyage was re-enacted in the massive television serial entitled *The Voyage of the Beagle*, it would appear that the script writers, for purposes unknown, added further falsifications of their own making, thus misleading the general public regarding what actually took place.

A) CAPTAIN FITZROY
Fitzroy was a very able and highly respected Commander of the "Beagle", who, after some initial uncertainty regarding Darwin's suitability as his close companion, made him as welcome as he possibly could. FitzRoy was inclined at times to be moody, which made it very difficult for Darwin. However, a genuine friendship was struck between them, which was maintained long after the voyage. Darwin, in his autobiography, said, "His character was in several respects one of the most noble which I have ever known". His original manuscript however continued "though tarnished by grave blemishes."[26p76]

Throughout the voyage, Darwin held to his Christian beliefs. On one occasion, when he quoted the Bible as a final authority on a particular point, he was surprised when some of the crew laughed in disbelief. Darwin has sometimes been portrayed as arguing violently in support of his evolutionary theories against the deeply held religious faith of the Captain. This is simply not true, for Darwin did not have any ideas on evolution until *after* the voyage. The only occasion when there was a sharp disagreement was over the question of slavery. Darwin felt he had lost his friendship with FitzRoy in view of the latter's outburst, but within a few hours FitzRoy had sent his apologies and Darwin continued to dine with the Captain.

B) GEOLOGY
It is very clear that Darwin's main interest throughout the voyage was Geology and *not* Zoology, for at each port of call he immediately tried to establish the age and history of the rock

formations he discovered. Indeed he complained that the collecting of the specimens of animals and plants interrupted his interest in the geology of an area. So inadequate was his knowledge of fossils that he rarely knew the names of any of those which he sent back to England, where their species were actually determined. He collected them as *geological* specimens rather than zoological. It was only some months after his return to England that he treated them with a great deal more attention in view of their possible implication in his theory of evolution for it was only then that he came to appreciate their relationship with the living forms.

C) ZOOLOGY

Darwin did collect numerous animals and birds for shipment back to England. In his Autobiography however he admitted:

> "...but from not being able to draw, and from not having sufficient anatomical knowledge, a great pile of MS [manuscripts] which I made during the voyage has proved almost useless."[2p62]

Huxley also commented that all his zeal and industry only resulted in a vast accumulation of useless manuscript.[1p92]

Similarly, during his time on the Galapagos Islands, he did not fully appreciate that the animals of islands so close together and of the same climate could nevertheless be so different.

It can therefore be seen that the image of Darwin's methodically collecting the evidence for his theory whilst he was on the voyage has no basis in fact whatsoever.

D) THE GALAPAGOS ISLANDS

It is this location where Darwin is said to have first seriously doubted the fixity of species. It is understandable that this should be accepted as a 'well attested fact', for Darwin himself claimed this.

In his autobiography he said -

> "During the voyage of the *Beagle* I had been deeply impressed...by the South American character of most of the productions of the Galapagos archipelago, and more especially by the manner in which they differ slightly on each island of the group....It was evident that such facts as these, as well as many others, could only be explained on the supposition that species could be modified; and the subject haunted me....I had always been much struck by such adaptations....After my return to England it appeared to me that by following the example of Lyell in Geology, and by collecting all facts which bore in any way on the variation of animals and plants under domestication and nature, some light might be thrown on the whole subject. My first notebook was opened in July 1837. I worked on truly Baconian principles, and without any theory collected facts on a wholesale scale,

more especially with respect to domesticated prod-
uctions, by printed enquiries, by conversation with
skilful breeders and gardeners, and by extensive reading"
[2p82].

Darwin clearly wished to imply that he was keen to return to
England in order to investigate the subject of species-variation. I
have quoted this passage at length as I will be commenting upon the
accuracy of Darwin's assertions later.

On an earlier page in his autobiography (p65), he wrote:

"Nor must I pass over the discovery of the singular
relations of the animals and plants inhabiting the several
islands of the Galapagos archipelago, and of all of them to
the inhabitants of South America".

Himmelfarb has noted that this passage was a later insertion in
the original manuscript and comments –

"It is apparent that, with the passage of time, the
events of the trip fixed themselves in his mind more and
more in the shape of a formula. This may account for the
fact that the same phrases repeat themselves with suspic-
ious exactness in his recollections, an exactness testifying
not so much to the trustworthiness of his memory as to the
rigid pattern imposed by memory on the events".

Himmelfarb examined Darwin's notebooks and letters *which he
made during his stay there* in order to ascertain precisely what were
Darwin's views regarding the animals on the Galapagos Islands. *His
notebooks give no hint of evolutionary thought.* Independant
confirmation of Himmelfarb's statement is provided by Professor
Gruber, an ardent admirer of Darwin, who spent a whole summer
examining Darwin's *Beagle* notes to find evidence of early ideas of
evolution in them. He admitted that "...there was little or nothing of
that sort..."[62pxiv]

Several pages of other notes however were found in which
Darwin refers to his observing the differing varieties, and concludes
that "...such facts would undermine the stability of species". The
editor of these notes, Nora Barlow (Darwin's grandaughter),
assuming they were made on the voyage, claimed they proved
"beyond doubt when these ideas crystallized". Himmelfarb, however, shows that the specimen numbers are out of sequence, whilst
the ink of the page numbers is identical to that of the sheet itself,
showing that they were written out in the order they appear. This
means that they are notes on bird specimens collected into groups
some time *after* the voyage, which, by inspection of Darwin's diary,
she dates as June 1838, a year and a half after his return.

It was seemingly towards the end of his time on the islands that he
was told by the inhabitants that the varieties of tortoises and other
animals and plants differed between the islands. By inspecting a

specimen they could tell the island from which it came. Yet, although he was taken on the voyage as the expedition's naturalist, and both he and his biographers credit him with "great powers of obvservatiion of detail", *he completely missed this fact himself.* Indeed, he did not appear to give it much attention, for he partially mixed the specimens from two islands, and did not gather a complete series of specimens from any one island. Whilst he was there, he did not think the species on different islands would be dissimilar, and his failure to collect evidence for this rankled him. Indeed, years later he admitted to Henslow, "I need not say that I collected blindly, and did not attempt to make complete series, but just took everything in flower blindly". Despite this admission, he claimed blandly in his autobiography:

> "...I think that I am superior to the common run of men in noticing things which easily escape attention, and in observing them carefully. My industry has been nearly as great as it could have been in the observation and collection of facts."![2p103]

Even with regard to his famous 'finches', he did not notice the various shapes of their beaks when he collected them, for this was only pointed out to him when his collection had been classified by an expert several months after he had returned to England.[62p299]

His lack of interest in the animals of the Galapagos Islands is shown in his correspondence. In a letter to Henslow he mentioned briefly: "I shall be very curious to know whether the Flora belongs to America or is peculiar" and "I paid also much attention to the birds which I suspect are very curious". He then immediately returned to the subject of the geology of the island, which was his main interest on the voyage. That it was the geological formations of the islands which impressed him more than the animal life that flourished there is supported by a letter which he wrote a few months later in which he described the islands as "that land of craters". In addition his notes on the geology of the islands were more than three times as long as those on the zoology.

His notebooks are particularly revealing for, amidst various unremarkable observations, he referred to the tortoises as "old-fashioned *antediluvian* animals" and wondered: "It will be very interesting to find from future comparison what district or '*centre of creation*' the organized beings of this archipelago must be attached".

Darwin used the word 'creation' several times in his notes and it is quite obvious that this is used in its normal sense i.e. that the species were created by God.

E) SOUTH AMERICA

Darwin's claim that it was whilst on the Galapagos Islands that he doubted the stability of species has been examined and has been shown to be quite false. He made a similar claim in the opening paragraph of the *Origins*, for he said he was "much struck with certain facts in the distribution of the organic beings inhabiting South America, and in the geological relations of the present to the past inhabitants of that Continent".

Himmelfarb has shown that this statement is likewise incorrect, for it is quite clear from his notes that he was only interested in the fossils from a *geological* standpoint, as he hoped that they would correlate the various strata of the South American Continent. It was not until Owen had identified the various fossils after Darwin had returned to England, that he related them to the present-day inhabitants.

From all this, it is certain that Darwin, throughout his voyage, and for some time after, was a *creationist*. Never at this stage did he seriously consider that one species might possibly evolve into another. Yet hardly any of his numerous biographers have ever given this fact its due prominence.

The questions which now arise are:
    When did he begin to consider that the evolution of one species from another may have taken place?
    Why should he have changed his mind?

The first of these is easy to answer. The second is not so easy, although there is every indication that he was strongly influenced by one person in particular.

CHAPTER 12

RETURN TO ENGLAND.

Darwin landed on 2nd October 1836, conscious that he had already attained some degree of fame. Henslow had received from Darwin a continuous flow of interesting specimens and fossils and was so impressed with his letters that he had some privately printed and circulated. Sedgwick, too, visited Charles' father and assured him that his son would take a place among the leading scientific men. On his return, Darwin was quickly introduced to the prominent scientists of the day, mainly through the efforts of Charles Lyell, who made his acquaintance.

Darwin spent the winter of 1836-7 at Cambridge. He was made Secretary of the Geological Society in February 1837 and resided in London until he moved to Downe, Kent, in 1842. He obtained admission into the prestigious Athanaeum club, where, it might be expected, Darwin would eagerly discuss his views with famous scientists of his day. He did indeed meet some leading scientists. However, Himmelfarb has noted that: "Historians, however, rather than scientists, seemed to be the staple of most of the dinner parties he attended...". One is left wondering why Darwin, as a scientist, should have preferred the company of those who write- or re-write- the history of the nations. Darwin also dined with various famous historians at Lord Stanhope's house a number of times and occasionally he had a meal with the old Earl Stanhope.

First notes on evolution

It was not until some ten months after his return to England, that there is any clear evidence that Darwin started to consider evolution as a viable possibility. In a notebook dated July 1837, he claims that in March of that year he was "greatly struck" by the South American fossils and the Galapagos species indicating that species were transmutable. Even this note is thought to have been added later, although the dating is approximately correct. This notebook was followed by several others up to October 1838. In 1842 he wrote a brief 'sketch' of the theory and in 1844 he wrote a more complete work for which he made arrangements for its publication should he die. In these early notebooks *virtually the whole of the theory of evolution is set out in considerable detail.* Thus his claim in both his autobiography quoted above (and a similar passage in the opening paragraphs of his *Origins*), that he "worked on true Baconian principles, and *without any theory*

collected facts on a wholesale scale..." is simply quite false, for in
these early notes he had developed the scheme of the theory in
considerable detail and long before he had collected many 'facts'
which would be published some fifteen years later.

Similarly, in a letter to Hooker in which he refers to his 1844
sketch on evolution, he said he had never ceased collecting facts
and now at last a "gleam of light" had come!

Darwin had clearly intended to give the impression that he had
been acting as an impartial student of scientific facts, when in
reality he was a wildly speculative theoriser who diligently searched
only for evidence which would support the ideas to which he was
already committed.

Himmelfarb comments:

> "Later, justifying the *Origin*, he defended the proce-
> dure of 'inventing a theory and seeing how many classes
> of facts the theory would explain.' - a more apt description
> of his method and of that of most scientists than the
> ritualistic cant about 'Baconian principles'".

The long gap

These first notes were made in 1837, some twenty-two years
before he published his *Origins*. The inference is usually drawn that
over this length of time he was painstakingly collecting the necessary
mass of evidence to support his contentions. Indeed, Darwin
claimed this in the opening paragraph of his *Origins* saying:

> "After five years' work I allowed myself to speculate
> on the subject, and drew up some short notes; these I
> enlarged in 1844 into a sketch of the conclusions which
> then seemed to me probable: *from that period to the
> present day I have steadily persued the same object.* I
> hope that I may be excused for entering on these personal
> details, as I give them to show that I have not been hasty
> in coming to a decision [!]".

He made a very similar statement in his autobiography, for he
wrote: "In July I opened my first notebook for facts in relation to the
Origin of Species, about which I had long reflected, *and never
ceased working for the next twenty years*" [2p68].

The truth, however, is quite otherwise. On his return to England,
Darwin was plunged into writing a section of the official *Journal* of
the voyage. Next he wrote on the subject of coral reefs. In 1842 he
wrote his brief first 'sketch' on the theory of evolution, followed by
his much more complete work in 1844. The period between,
however, was almost fully occupied by his writing on the *Geology of
Volcanic Islands*.

He worked for two years on the *Geology of South America*,
intending to start on his 'species work' as soon as he had studied

briefly some barnacles. This work on barnacles, however, continued to grow, taking him eight years to complete! The result was an enormous work in four volumes. When it was completed, Darwin had to dispose of some 10,000 barnacles.

[An amusing tale is related about Darwin's children. Under the impression that all fathers followed the same occupation, they asked a neighbour's children about their father, "Where does he do his barnacles then?"]

I mention all these papers he wrote to show that, although he corresponded with various people on the subject of evolution, Darwin actually spent only a comparatively small portion of the time up to 1854 on the mutability of species. His output on this consisted solely of his notebooks started in 1837 and his 'sketches' of 1842 and 1844.

Confirmation of this is contained in a letter he wrote to Hooker in 1854 in which he said "I have been frittering away my time... sending ten thousand barnacles out of the house all over the world. But I shall now in a day or two *begin to look over my old notes on species.*"[2p395]

Thus it was not until 1856, being urged by Lyell, that he started a full time study of the 'mutability of species'. This was intended to be a very much larger work than the *Origins* and was half completed by June 1858. It was then that Darwin was stunned by the reception of a manuscript from Alfred Wallace which set out exactly the same lines of the theory upon which Darwin himself was working.

CHAPTER 13

ALFRED WALLACE

ALFRED RUSSELL WALLACE (1823-1913) was particularly interested in such subjects as spiritualism, psychical research, mesmerism and phrenology. He was also, like Darwin, interested in the problem of species and a self-taught amateur biologist when, from 1848 to 1852, he spent four years with H.W. Bates in the Amazon Basin collecting specimens for sale. In 1854 he left again for the Malay Archipelego where he stayed for eight years making further collections. Whilst he was there he wrote a paper *On the Law which has Regulated the Introduction of New Species* which was printed in a scientific journal. Lyell noticed this and drew Darwin's attention to it. He appears to have warned Darwin that he should hurry up and finish his book or Wallace may produce a work on the theory before him. Darwin was to regret not heeding Lyell's advice.

Wallace's ideas were very similar to Darwin's, but he did not propose any mechanism whereby new species could arise. He wrote to Darwin who, in his reply, clearly told him that he was already working on the subject and expected to publish it as a major work in two years' time. Wallace, however, ignored this 'warning' and started to write a book on the subject. It was then that Wallace remembered Malthus' paper. He had read this some years before and it now suggested to him the idea that the improvement of species was brought about by the 'survival of the fittest'. Accordingly he wrote out his thoughts in a rush and sent them to Darwin for his opinion, and then to be sent on to Lyell. Darwin realized that his theory was exactly the same as Wallace's and appealed to Lyell for advice. Lyell proposed the paper should be read together with one by Darwin at a meeting of the Linnean Society, which duly took place on 1st July 1858.

One surprising feature of this presentation was the way in which these papers were *completely ignored*! They were read in place of a paper concerning the *immutability* of species that was withdrawn at the last moment. Yet there was *no discussion* on these two papers by the thirty members who had assembled to hear views quite the opposite!

One member, S. Houghton, later assured the Dublin Geological Society that had it not been for the authority under whose auspices it appeared (Lyell read one paper, Hooker the other), it would not have been worthy of remark. Similarly, no review appeared in any contemporary journals. Neither did a single comment appear in the annual review of events by the Linnean Society. Such a cool reception would confirm that amongst serious and able scientists of the day, the general fixity of species was so well accepted that vague theorizing as proposed by Darwin and Wallace was not taken particularly seriously. It was not until the appearance of Darwin's *Origins*, with its massive parade of 'facts', was the subject given much greater publicity and eventually a change in thinking took place.

Following this meeting, Darwin began the writing of his major work which was published a year later.

Two strange aspects

In considering this incident of Wallace's letters to Darwin, there are two strange aspects.

First, both Wallace and Darwin claimed that it was Malthus' essay which suddenly shone like a bright light on the theory and provided the vital piece of the jig-saw necessary to complete it. Yet Malthus' paper was written in 1798 *some sixty years earlier* and must have been well known to the historians and scientists with whom Wallace and Darwin mixed. The "struggle for existence" had been postulated for many years and it did not really require Malthus' paper to propose this as a fresh idea. Furthermore, as we have shown, Malthus predicted a degradation of the (human) species, *not* the improvement of fit survivors!

Secondly, the timing of Wallace's letters appears to be very fortuitous in the production of Darwin's *Origin*.

Darwin had returned from his voyage in the *Beagle* eighteen years and had still not started writing a full work on 'species', when Wallace's first paper was published in 1854. Lyell drew Darwin's attention to it and urged him to start, which Darwin did in 1856. This enormous work was only partly completed when Wallace, four years after receiving Darwin's 'warning', ignored it and sent his famous letter in 1858, in which he too claimed inspiration from Malthus. He asked Darwin to forward his paper on to Lyell, who proposed a joint reading of their papers, which duly took place.

Darwin was now galvanized into considerable activity and wrote his abstract *Origins*, a lengthy work, in just about twelve months! Wallace then quietly dropped out of the scene.

It is rather interesting that the effect of both of Wallace's articles was to goad Darwin into action. For this purpose they could not have been timed more strategically. This was presumably purely coincidental, but one cannot help wondering whether the sequence of events might have been discreetly engineered for this precise purpose.

Examination of Wallace's autobiography does not provide any direct evidence in support of this idea. Yet, interestingly, it does raise a parallel question regarding Wallace's reticence concerning his first acquaintance with Darwin.

In his autobiography, Wallace gives numerous personal details and anecdotes of people that he met and knew in the course of his active life. When he sent his letter to Darwin from Malaya, requesting that it should be forwarded to Sir Charles Lyell if he approved of it, one has the impression that he was well acquainted with both of them. Such however does not seem to be the case.

His return from South America marked the beginning of his acceptance into the circle of naturalist and prominent authorities on the London area. One would therefore expect him to describe his first meeting with such people in some detail. Yet in dealing with this important period in his life he glosses over it in only six pages. He mentions that he was well known to the authorities of the Zoological and Entomological Societies and was welcome to their meetings. He presented a paper at one of the meetings and attended a lecture by Huxley. Yet he was clearly a member of the 'establishment' for he relates how he obtained a free passage to the Far East:

> "As the journey to the East was an expensive one, *I
> was advised to try and get a free passage in some
> Government ship*[!]....I had made the acquaintance of
> Sir Roderick Murchison...one of the most accesible and
> kindly men of science. On calling upon him and stating
> my wishes, he at once agreed to make application on my
> behalf...as he he was personally known to many members
> of the Government and had great influence with them."
> [51p326]

When the first ship changed its destination, Wallace called on Murchison again, who quickly obtained a first-class ticket for him on the next sailing of a P&O steamer.

It was after he had sent his first paper on species for publication, that he makes his first reference to Darwin. He gives no indication of any previous contact with him but in the course of his narrative casually remarks:

> "...I had in a letter to Darwin expressed my surprise
> that no notice appeared to have been taken of my paper, to
> which he replied that both Sir Charles Lyell and Mr.
> Edward Blyth...specially called his attention to it."
> [51p355]

Darwin's reply is quite formal for he begins "My dear sir...", from which it can be presumed that he had never met Wallace. Wallace became a close friend of both Lyell and Darwin *after* his return from Malaya but makes no mention of them before he went there. In his autobiography he provides many details about the various famous personages he met, yet with regard to his first contact with one of the foremost men of his day he is surprisingly silent. One is therefore left wondering what (or who?) prompted Wallace to write to Darwin - a well known scientific personality whom he had never met and with whom he had never corresponded before.

CHAPTER 14

GATHERING THE EVIDENCE

Darwin used a variety of sources for his evidence in building up his case. He read numerous scientific papers, but much of the information he actually used came from discussions and correspondence with scientific colleagues and particularly with "skilful breeders and gardeners". He maintained a vast correspondence with various informants asking such strange questions as: "At what age do nestling pigeons have their tail feathers sufficiently developed to be counted", and: "Did you ever see a black greyhound (or any sub-breed) with tan feet and a tan coloured spot over inner corner of each eye? I want such a case, and such *must* [emphasis his] exist because theory tells me it ought."

As a result his notebook contained such statements as:

"Strong odour of negroes - a point of real repugnance."

"My father says, on authority of Mr. Wynne, the bitch's offspring is affected by previous marriages with impure breed."

"The cat had its tail cut off at Shrewsbury, and its kittens had all short tails."

In a letter to Huxley he described how he sat one evening in a gin-palace in the Borough with a number of pigeon-fanciers and he recounted some of the talk about crossing various breeds and the problems involved. He was impressed by the general shaking of heads when the crossing of two breeds was mentioned in an effort to improve the stock, and he commented: "*All this was brought home far more vividly than by pages of mere statement and Co*"[3p282].

From such sources did Darwin glean many of his 'facts'. However, as Himmelfarb noted:

"All too often, however, the practical wisdom of these men is revealed to be more the fruit of prejudice than of mature experience. And unfortunately Darwin could not distinguish between the two...."

Darwin appears to have been very naive even as a child - a characteristic which he never seems to have entirely outgrown in his mature years. In his autobiography he almost proudly claims:

"I am not very sceptical, - a frame of mind which I believe to be injurious to the progress of science."
[2p104]

Darwin appears to be confusing the carrying out of unusual experiments, which is a perfectly reasonable procedure, with the acceptance of information which may be second or third hand - a practice to which he was particularly susceptible.

Darwin's instruments

Darwin's very casual approach in acquiring information extended to his 'instruments' (if such they can be called). The ruler he used was in common use by the rest of the household. He preferred the simple microscope to a compound one. His seven-foot rod for measuring plants was graduated by the village carpenter. His scales for weighing were of poor quality, whilst his chemical balance was the one he used when he and his brother made experiments in their garden shed when he was sixteen years old. Volume measurements were made with great care (as were all his measurements), using an apothecary's glass which was roughly made and badly graduated, and he was quite astonished when informed that his two micrometers were giving quite different readings! As if such a catalogue were not enough, he converted from inches to millimeters using a figure obtained from an old book which was found to be inaccurate! [2p148].

Such poor instruments would not of course invalidate his results, but do indicate that he lacked the sense of precision which might be expected in a researcher even of his generation. Indeed, when this is considered together with the anecdotal form in which he obtained much of his information, a picture is gained, not of a professional scientist but of an amateur whose standard was well below that of many other amateurs of his day. Therefore any suggestion that Darwin was a real 'scientist', in the accepted meaning and status which the word implies, is quite unjustified.

CHAPTER 15

THE PUBLICATION
OF THE *ORIGINS*

When Darwin completed his book *The Origin of Species*, the publisher, John Murray, was induced by Lyell to publish it. Darwin, however, suspected that it was against Murray's better judgment. Murray agreed to publish it after he had received the chapter headings, but when he had read the whole manuscript he had great reservations, not only about its scientific value, but also its commercial viability. An amateur geologist himself, he considered the theory "as absurd as though one should contemplate a fruitful union between a poker and a rabbit." He asked for an opinion from the editor of the highly-respected *Quarterly Review*, who advised Darwin not to deal in such a controversial subject and write a book on pigeons which was bound to make him famous! Murray asked for a further opinion from a lawyer friend who suggested that it was "probably beyond the comprehension of any scientific man living". Nevertheless he induced Murray to print one thousand copies rather than the five hundred he intended.

Vacillation

Darwin's attitude at this time is interesting. With the amount of research he professed to have carried out and the array of 'facts' he provided, one might have expected him to be reasonably confident in the correctness of 'his' theory. Darwin, however, vacillated about its success and wrote to Murray giving him complete freedom to withdraw from publishing the book, saying: "...though I shall be a little disappointed, I shall in no way be injured." In a letter to Hooker he asked him "not to say to anyone that I thought my book on Species would be fairly popular, and have a fairly remunerative sale...for if it prove a dead failure, it would make me the more ridiculous" [3p157]. Even in the correction of the proofs, Darwin rewrote so much of the book that he offered to pay Murray for the heavy cost of the amendments.

Just how unconvinced Darwin was by his own theory is revealed in two letters he wrote about a year after the publication of the *Origins*. The first of these was to Lyell in which he said:

"For myself, also, I rejoice profoundly; for thinking of so many cases of men persuing an illusion for years, often and often a cold shudder has run through me, and I have asked myself whether I may have not devoted myself to a phantasy."[3p229]

The second was to Huxley in which he admitted "Exactly fifteen months ago, when I put pen to paper for this volume, I had awful misgivings; and I thought perhaps I had deluded myelf as so many have done..."[3p232]

These passages show just how very uncertain Darwin was of the truth of his theory which he nevertheless confidently propounded in his book.

THE 'SELL OUT'

Under such doubtful predictions was the *Origins* eventually published on the 24th November 1859. In describing this event, historians of evolution invariably add the phrase that "it was completely sold out on its first day of publication". Most readers would naturally imagine that large numbers of the scientifically-inclined members of the general public, who had long been awaiting the publication of this book, promptly crowded into the bookshops, clamouring to secure a copy of this new revolutionary work.

In fact no such demand ever arose in this way, for what actually took place was that *a total printing of only 1,250 books was fully subscribed by the various dealers and book agents at Murray's annual sale a few days before the publication date.* This was a perfectly normal and unremarkable transaction for many books published in the trade. Indeed the first few sales were quite unexceptional, the book appearing in sixth place in Murray's advertising list. It appeared in a similarly inconspicuous position when a second edition was published in January 1860.

Thus the claim that it "sold out on the first day" is only technically correct and does not warrant the implication that it was as greatly in demand which such a plain statement suggests. It is nevertheless so frequently repeated by evolutionists that it is unlikely that such a statement would be otherwise than accepted at face value. This is just one further example of how the public can be misled by a slanting of the facts. In this case the purpose is to give the impression that *Origins* was eagerly awaited and highly acclaimed on the very first day it appeared.

CHAPTER 16

A CRITICAL REVIEW OF "ORIGINS"

We will now examine this major work of Darwin's, which is rightly regarded as a watershed in the whole history of the rise of evolution. Before 1859, the theory received scant regard from the professional scientists. After the publication of *Origins*, evolution eventually became the new fashionable theory in many sciences, as well as forming the chief topic of conversation at many dinner parties. Eventually it dominated not only the physical sciences but also the social and political fields.

What was there in this book which caused such a revolution in men's thinking? Did Darwin really assemble and parade before his readers such a range of detailed, documented and unanswerable facts that any normal person with a grain of common sense would be forced to accept his theory as indisputably proven? I fear not. Indeed, the absence of irrefutable facts and the high degree of hypothetical theorizing is one of the most striking features of this large book. There are many criticisms which could be made of this work and we will examine the more important ones.

A) INADEQUACY OF FACTS

What evidence did Darwin actually produce to support his theory? As has already been noted, he relied very heavily indeed upon the 'stories' (for they were sometimes little more than this), communicated by a number of breeders and scientists both amateur and professional. One has the sense when reading his book that and most inadequate anecdote has been recorded for future use in order to bolster his case. He used such information in two ways.

Firstly, in wishing to prove a particular point, he would often quote only *one* example in support and thereby consider this as ample justification. To adequately prove even one small scientific point usually requires several examples of fully corroborated evidence.

Secondly, he would quote examples in one area under consideration and then apply it to quite a different situation, without alerting his reader that the circumstances did not apply in the broader field. For example, in the discussion on the problem of why hybrid crosses between different species are sterile, he quotes various instances to illustrate his arguments. His line of reasoning, however, is so confusing as to be almost impossible to follow and I will be giving an example from this chapter later. He seems to have almost deliberately confused his readers, for he eventually contended at the end of the chapter that, although we do not know why crosses should be sterile, "yet the facts given in this chapter do not seem to me to be opposed to the belief that species aboriginally existed as varieties".

Now the important point is that, although throughout the whole of this chapter he continually uses words such as 'hybrid', 'sterility', 'species', and so forth, in practice he deals almost entirely with experiments on plants, where hybrids and crosses are more frequently obtained, because of their particular genetic structure. When he deals with animal hybrids he spends only two pages out of a chapter of thirty pages and begins with the admission "Although I know of hardly any thoroughly well-authenticated cases of perfectly fertile hybrid animals, I have reason to believe that ..." and then proceeds to give a few examples. He then reverts to the question of hybrids of various plants. Thus the reader, having been led to assume that hybrids *are* possible in the plant world, would naturally infer that this would apply to the animal world also where the evidence is much less convincing.

It is interesting that Darwin's brother was aware that the facts were being distorted, for having read the *Origins*, he wrote to Charles saying:

"...the a priori reasoning is so entirely satisfactory to me that if the facts won't fit in, why so much the worse for facts is my feeling" [3p233].

B) VAGUE AUTHORITIES

It is not usually appreciated that Darwin in his Introduction admitted that his book was only an "abstract" of his ideas, that he could not "here give references and authorities" for his statements, for he "must trust to the reader for reposing some confidence in my accuracy". Acknowledging all this, he hoped "in a future work" to publish "in detail all the facts, with references". Needless to say, no such work was ever published, *and neither did he ever start work on one*, nor did he return to the larger work begun in 1856. Thus he convinced his readers that there was a whole body of conclusive facts behind his theory which he had merely to set down in order to complete his case!

Although he often acknowledged some eminent person as the source of his information, it generally appears that he received it either by word of mouth or by personal letter. Indeed, *he almost never quoted a single scientific publication in support of his theory* but merely gave the expert's name. Thus he made it extremely difficult to check any of his facts by examining the scientific journals of his day. This alone justifies the label 'unscientific' being applied to his publication.

One excuse which he used for not providing full details for the reader to follow up was that the omission was necessary "due to lack of space". Yet the overall work consists of more than 200,000 words and to have given detailed references would have added only a few more pages.

C) HYPOTHETICAL NATURE

The very large number of assumptions and hypotheses which Darwin employed in support of his ideas is very evident and is generally conceded by his supporters. Such phrases as:

"probably",

"may perhaps ",

"might have been",

"but if this has occurred",

"it is conceivable that",

"I can see no insuperable difficulty in believing that",

and similar apologetic expressions are liberally sprinkled throughout the work. Indeed, on some occasions where examples failed him completely, or he wished to illustrate a particular 'principle', he says: "Let us take an imaginary example..."

One subterfuge (and I use this word advisedly) he used when faced with a difficulty was to suggest a possible series of events which would lead to the result he desired thus overcoming the objection. Having delivered this he would then *assume* that he had satisfactorily proved his point! It is a far cry from suggesting what *might* have happened to proving that it *did* happen. This invention of "explanatory theories" is often used by defenders of evolution today.

In a similar fashion, he would occasionaly suggest a possible explanation in one chapter, and then later when referring to this same subject, he would say: "As we have already demonstrated previously...". Thus, what was a hypothesis is later claimed as a proven fact.

D) DEVIOUS ARGUMENTATION

When a theory is presented, it should be accompanied by an abundance of facts. Such is not the case with Darwin's *Origins*. Although he provided a fair number of unconvincing single examples, in certain crucial areas he spent much space in *trying to suggest possible explanations of why the facts he needs are not found*. This is particularly noticeable when dealing with the large gaps in the fossil record. He took the whole of one chapter in an attempt to explain these gaps by a series of assumptions and unverified statements, in which he claimed that the geological strata provided only a very fragmentary record of past ages. He suggested that the links between any two particular species were necessarily few in number and rapidly died out as a result of pressure from the two larger groups. Yet he did not attempt to consider why the first different offspring, which would only have a *very* small advantage over the parent stock, was not similarly eliminated as a result of pressure of the parent stock and other numerous species.

He suggested that new species rapidly developed in remote parts of the world where the strata had not been thoroughly examined, and that these had appeared later in other areas in large numbers as a populus new species. So on the one hand he consistently presented the picture that evolution had taken place very slowly over millions of years, yet on the other hand, when the facts were missing, he proposed that new species must have developed more rapidly in remote parts of the world where the evidence had not yet been found, or had been eroded away.

Another aspect of Darwin's deviousness is the difficulty experienced in attempting to follow his arguments. At times he switched from the main topic to side-issues. Then he would give an extensive account of his findings and experiments regarding a further problem, finally concluding that all these points he had considered supported his theory.

Indeed, I have a strong suspicion that his involved line of reasoning and the very length of the book were both deliberately contrived features. By this means the reader would be 'blinded with science' and, at the same time, the inherent weaknesses of the arguments are made more difficult to perceive.

On this aspect, Himmelfarb comments:

"The points were so intricately argued that to follow them at all required considerable patience and concentratiion — an expenditure of effort which was itself conducive to acquiescence" [1p288].

Darwin's complicated way of arguing was recognized by his own sons when writing his biography, for they noted:

"...his strange habit of containing two contradictory or conflicting thoughts in his mind at the same time and trying to express them both simultaneously, or introducing his quali- fications before his main statement" [1p388].

Irvin also comments:

"He was a slow reader, particularly in foreign languages. He could not draw. He was clumsy and awkward with his hands, and despite his interest and belief in experiment, he was in some ways oddly careless and inefficient....'He used to say of himself that he was not quick enough to hold an argument with anyone,' and his conversation was an adventure of parantheses within parantheses which often produced a stammer and sometimes terminated in unintelligibility and syntactical disaster. He wrote fairly clear and interesting English only by slowly and painfully improving the impossible, and when he took pen in hand laughingly grumbled that 'if a bad arrangement of a sentence was possible, he should be sure to adopt it' " [16p56].

To give the reader just one example of Darwin's tortuous style of writing, I quote a passage (which is one long sentence) from chapter V in which he deals with variations which occur within species:

"In these remarks we have referred to special parts or organs being still variable, because they have recently varied and thus come to differ, but we have also seen in the second chapter that the same principle applies to the whole individual; for in a district where many species of a genus are found - that is, where there has been much former variation and differentiation, or where the manufactory of new specific forms has been actively at work - in that district and amongst these species, we now find, on an average, most varieties"[19p155].

E) CIRCULAR ARGUMENTS

The use of fossils in the rock strata to support the theory of evolution has been criticized for being a circular argument, i.e. the fossils are assumed to be arranged in an evolutionary sequence which is then used to date the strata. The strata are then used to date the fossils and the evolutionary sequence obtained is then held to prove the theory!

In a not dissimilar fashion, Darwin also used arguments which are now seen to be circular. For example, the catch phrase "survival of the fittest" claims that those who are "the most fit" survive to leave most offspring. But it is impossible to define a fit from an unfit organism - given the complex environmental conditions in which they live. Thus it can only be said that those who survive to leave most offspring are those who leave most offspring!

This whole subject of circular reasoning and faulty logic, which are now apparent in many of Darwin's arguments, has been thoroughly exposed by Norman Macbeth in his book *Darwin Retried* [37]. Macbeth, a retired lawyer, not only exposes the inadequacies of Darwin's logic, but of those of the Neo-Darwinists and other modern variations of evolutionary theory. His book is very damaging to the whole basis of the arguments used to support evolution. It was therefore surprising to hear him say in a television programme that nevertheless he considered that evolution had taken place. Having demolished the theory so thoroughly, it is far from clear what he found so convincing in the evidence which was left.

PROFESSOR THOMPSON'S INTRODUCTION

The weaknesses of Darwin's arguments have been much criticized. One of the most notable critics is Professor W.R. Thompson, F.R.S. He wrote an introduction for the 1956 reissue of the *Origin of Species* by J.M. Dent and Sons, at which time he was Director of the Commonwealth Institute of Biological Control. Having warned the publishers that he would not be writing a 'hymn of praise' to Darwin, to which they did not demur, he proceeded to criticize the very basis of Darwin's theories and his arguments. In his introduction he commented:

> "The success of Darwinism was accompanied by a decline in scientific integrity. This is already evident in the reckless statements of Haeckel and in the shifty, devious and histrionic argumentation of T.H. Huxley",

and:

> "To establish the continuity required by theory, historical arguments are invoked even though historical evidence is lacking. Thus are engendered those fragile towers of hypotheses based on hypotheses, where fact and fiction intermingle in an inextricable confusion".[59]

It is interesting that when the publishers reprinted the *Origins* in 1971 they invited an evolutionist, L. Harrison Matthews, to provide the introduction. In this he candidly admits that "Much of Professor Thompson's criticism of Darwin's text is unanswerable." and continues to make a number of points which are very damaging to the evolutionist's cause. He notes:

> "[Adequate] proof has never been produced, though a few not entirely convincing examples are claimed to have been found[He later quotes the case of the Peppered Moth]. The fact of evolution is the backbone of biology, and biology is thus in the peculiar position of being a science founded on an unproved theory - is it then a science or a faith ? Belief in the theory is thus exactly parallel to the belief in special creation - both are concepts which believers know to be true but neither, up to the present, has been capable of proof."[19]

SUMMARY

To subject such an influential book as the *Origins* to such heavy criticism may well be regarded by some as unjustified. This does raise the question of how many people have actually read Darwin's book. The way in which it is rarely quoted by scientists as a source reference in their technical papers suggests that it was used not so much as a researcher's text book but more as a rallying point and propaganda weapon for supporters of evolution.

To those who consider these criticisms as unwarranted but who have never read the book, I would suggest that they open a copy of the *Origins* at random, and in the light of my comments above make a critical assessment of the work. I trust that if nothing else he will agree that, whilst Darwin lacked the ability to convince his readers by presenting irrefutable facts and arguments, he was a past master in the art of sophistry — the dictionary definition of which is "to use false arguments intending to deceive".

CHAPTER 17

THE RECEPTION OF *ORIGINS*

Darwin's friends

The reactions of Darwin's three close friends to the published theory are interesting.

Hooker appears to have fully accepted the theory and was one of the first to write a book which referred to the theory as providing an explanation of the relationship between various plants and their environment etc.

Lyell, somewhat characteristically, vacillated so much that Darwin despaired at times of ever receiving his full approval.

Huxley, at first cautious when he read some preliminary proofs, found the complete book far more convincing. Nevertheless, he still had many reservations. He was well aware of the great weaknesses in the theory and, like Lyell, never claimed that he was completely convinced. In giving any lecture he would often begin by saying that he only looked upon evolution as a "working hypothesis", thus safeguarding his future status should the theory later collapse. Then he would proceed to support it and scourge the opposition in such a spectacular fashion that his preliminary cautions would have been quickly forgotten. In reading some of his papers, it is clear that he had a vivid imagination, a biting sarcasm and a brilliant use of classical metaphors. However, one is left with a feeling that he delighted in the public display of his powerful delivery, finding in addition that such attacks were the best means of defence.

The general reception

Although the sales of Darwin's book were not at first as phenomenal as popular writers would have us believe, the importance of this controversial theory was quickly realized by both the churchmen and scientists of the day. Naturally the former recognized that the theory provided a means of removing the necessity of an all-powerful, creator God as given in the Old Testament. Although a few churchmen quickly joined the evolutionary camp, most of them were strongly antagonistic. Darwin expected their opposition, and that of most scientists, but what baffled him was that the scientists rejected the theory not just because it lacked convincing proof but because many of them held deep religious beliefs.

THE BRITISH ASSOCIATION MEETING

In many reviews, the book was heavily criticized, but the first 'success' claimed by evolutionists was the British Association meeting at Oxford in June 1860. Historians invariably refer to this meeting and quote Huxley's 'brilliant' riposte to a personal thrust by Wilberforce. The Bishop towards the end of his speech is said to have turned to Huxley and asked "whether it was through his grandfather or his grandmother that he claimed descent from a monkey?" Huxley, in his reply, said that he was not ashamed of having a monkey as an ancestor but would be ashamed of having as an ancestor a man who used his abilities in a sphere of science with which he had no real acquaintance and used aimless rhetoric in an appeal to religious prejudice.

It was this reply which it is insinuated 'won the day' for the evolutionist. It appears with great regularity in accounts of the story of evolution with the inference that this incident marked the point when the tide began to turn in favour of the theory.

Although there is no copy of what Wilberforce said at the Meeting, his main crticisms of the theory can be obtained from the review of Darwin's *Origins* which he wrote for the *Quarterly Review* July 1860. When Darwin read this he admitted:

"It is uncommonly clever; it picks out with skill all the most conjectural parts, and brings forward quite well all the difficulties. It quizzes me quite spendidly by quoting the 'Anti-Jacobin' versus my Grandfather."[3 p324]

What really caused the consternation in the meeting was not that Huxley had scored a resounding victory in the debate by proving Wilberforce wrong on any scientific basis but simply that he had been so direspectful to a Bishop - in those days an unheard of thing.

It must not be forgotten that there were no reports of the speeches and no voting at the end of the discussions, whilst the main sources of information are those provided by letters and comments by Darwin's followers. They would naturally be eager to seize upon the noisy uproar generated by a massed group of undergraduates in one part of the room as an acclaim of 'victory' for their side. Because of this it is sometimes implied that Huxley's speech convinced the majority of those present. This is contradicted however by a reference to Huxley's followers, who "...recalled the 'looks of bitter hatred' bestowed on them as they passed through *the still predominantly hostile crowd* after his speech" [1 p241].

Darwin's views created such sharp divisions that no account of the meeting could be considered unbiased. However, J.R. Lucas in a

Fig.9. T.H. Huxley in 1857
"I am sharpening up my claws
and beak in readiness"

Fig.10. Bishop Samuel Wilberforce

Fig.11. Charles Lyell in 1885

Fig.12. Alfred Russell Wallace

letter in*Nature* [67] has claimed that the actual progress of the debate was quite different to the popular version. Firstly he says that two journalists who were present did *not* consider that Huxley had scored a great victory, whilst Hooker felt that it was *his* reply which really made a case for Darwinism, with which Lyell appears to have agreed. Secondly, regarding Wilberforces famous question to Huxley he says:

> "It is doubtful that Wilberforce asked Huxley whether he was descended from an ape. It makes a good story, but Wilberforce had used the *first* [emphasis his] person plural in his review, and the use of the first person is borne out by Wilberforce's biography and one - admittedly late - account. What Wilberforce may have asked Huxley was where he drew the line between human desendents and ape-like ancestors......Huxley, however, *was ready to answer the question he had not been asked.* Three months earlier, in the April issue of the *Westminster Review*, he had accused the critics of Darwin of making him out to be no better than an ape himself, and since Wilberforce was now criticising him for being a darwinian, he must be calling him an ape too."

There is a hint that this account is correct in a letter which Huxley wrote to a friend about the debate. He said:

> "[I] had been unable to discover a new fact or a new argument in it - except indeed the question raised as to my personal predelictions in the matter of ancestry...*If* then, said I *the question is put to me* [emphasis mine] would I rather have a miserable ape for a grandfather...I unhesitatingly affirm my preference for the ape."[50p69]

Thus, it would seem that Wilberforce did *not* try to ridicule Huxley about his ancestry, but Huxley was prepared to make a very personal attack upon the Bishop on the slightest excuse. If this is correct, then it gives a totally different picture of what actually happened at this famous meeting of the British Association.

Samuel Wilberforce

As is so frequently the case, the overpraised heroes of one generation are found by succeeding generations to have feet of clay. The obverse of this process is that sometimes those who are spurned and mocked by their contemporaries may subsequently be found to be nearer the truth than is appreciated at the time. Such seems to be the case of Samuel Wilberforce.

The son of the famous William Wilberforce, who promoted the freedom of slaves, Samuel rose to become the Bishop of Oxford. He is invariably held up - by insinuation if not otherwise - as typifying the bigotted reactionary churchmen who tried vainly to stem the irresistible progress of 'science' by making a personal attack upon Huxley. This assessment, however, is now seen as being quite unjust.

He obtained a First in Mathematics and was recognized as being a persuasive and able debater in various fields and was a vice-president of the British Association. Before he spoke at the B.A. meeting he had prepared himself with all the weaknesses of Darwin's arguments, having apparently been briefed by Owen, the most outstanding natural scientist of the day.

An interesting review of the debate and an appreciation of Wilberforce's character appeared in an article in the *New Scientist* which concluded:

"Did the defeat embitter him? His attitude is sometimes portrayed as pompous and rather ill-humoured, but there is little evidence for this and an entirely different picture comes from his delightful portrait of A.E. Knox. Knox was a naturalist who eulogised the joys of fishing in his *Autumns on the Spey*, published in 1872. Wilberforce wrote a long and friendly review of the book, and he showed his appreciation of Knox's enthusiasm for his sport with a charming idea surely recalling his question to Huxley of twelve years previous. 'If Mr. Darwin's theory should ever be established', he wrote, 'there can be no doubt that Mr. Knox will be found to have descended, not from any prick-eared, tree-inhabiting monkey, but probably after the fewest interstitial gradations from some grand and venerable heron.' Perhaps the years had mellowed him but equally, perhaps this was his natural style, and he wasn't such a dinosaur after all."
[61]

Huxley himself later admitted that the Bishop bore him no ill will. Just how ambitious Huxley was can be judged from his purpose in attending a meeting of the British Association several years earlier in 1851. He confided to his fiancee:

"Anyone who conceives that I went down from any especial interest in the progress of science makes a great mistake. My journey was altogether a matter of diplomacy."[50p11]

Huxley had intended to leave Oxford but was persuaded by his friends to attend the Bishop's talk and he may well have seen the opportunity which it presented. It is therefore not impossible that Huxley may well have determined in advance to achieve notoriety for himself at this meeting. This would be sufficient to explain his unwarranted personal attack upon Wilberforce, who additionaly symbolised the Church Establishment which Huxley disliked so intently.

As often happens, the 'outcome' of debates are not so much a matter of the rightness or otherwise of the case but of the performance of the debaters involved and the subsequent reports. It seems likely that Wilberforce did *not* make a personal attack on Huxley, and this would certainly be in keeping with his gentlemanly nature. Despite this, it is certain that Huxley labelled the Bishop's lecture as "empty rhetoric" and then turned upon him with a quite unwarranted personal rejoinder that was far more wounding than was called for. The most noteworthy feature of Huxley's retort is its blunt rudeness towards a kindly and highly respected Church dignitary. It is *this* factor which shocked the audience and earned for his speech the very cheap publicity which it has ever since enjoyed.

The theory spreads

With the whole subject now entering the sphere of public debate, gradually support for the theory grew and one by one, over a period of years, various institutions and scientific bodies signified their approval. Today it is accepted as a 'fact' and forms the premise of so many of the scientific and sociological papers which are published. What Huxley forsaw has come to pass, for he said:

"History warns us...that it is the customary fate of new truths to begin as heresies, and end as superstitions; and, as matters now stand, it is hardly rash to anticipate that, in another twenty years, the new generation, educated under the influences of the present day, will be in danger of accepting the main doctrines of the *Origin of Species*, with as little reflection, and it may be with as little justification as so many of our contemporaries, twenty years ago rejected them....Whenever a doctrine claims our assent we should reply, Take it if you can compel it" [12p100].

It would seem that Huxley's own warning has gone unheeded.

CHAPTER 18

THE ACCEPTANCE OF THE THEORY

The fact that the evolution was only a hypothesis was grudgingly acknowledged by Darwin and his more reasonable supporters. As the years passed, however, it became an accepted 'fact', until today, to even question its validity is to invite ridicule and the accusation of scientific heresy.

The question must therefore be asked how such an unproven scientific theory came to be accepted by an increasing number of scientists so that today it provides the basis for all the most influential philosophies of life in the Western world.

The answer is found in the concurrence of various seemingly unrelated facets of life at the time when Darwin produced his book. Clark, in his book *Darwin Before and After*, gives an excellent survey of the rise of evolution and I am indebted to this work for some of the aspects which will now be considered.

1) THE 'SCIENTIFIC' VENEER
As was generally acknowledged, the ideas of evolution had been raised long before Darwin's time, but had been rejected by serious scientists, as the evidence quoted in its support was very weak. Darwin's achievement was to so arrange all the available indications (I will not call them 'facts') that evolution *might* have taken place, and present them with such subtle arguments and examples that they were invested with much greater importance than they actually warranted. In this way he presented the case for evolution with a scientific 'veneer' and thereby made it more readily acceptable to the rising intellectuals of his day who were fascinated by the dawn of the scientific era.

2) THE GREAT BIOLOGICAL 'LAW'
By the mid nineteenth century the physical scientists had discovered a number of the basic laws which govern the universe. Galaxies had been discovered and spectral analyses of stars had been made. Chemists were discovering the atomic structure of compounds. Faraday was investigating the laws of electricity, whilst Clerk Maxwell had discovered the mathematical relationship between electricity and magnetism. These were just some of the fascinating findings of the physical sciences which were beginning to have a social impact.

For the natural sciences, however, biologists, zoologists and botanists were discovering a vast number of facts *but no general 'law' which would combine any of them into a coherent pattern* in the same way as was the case in the physical sciences. Naturalists suffered from a sense of frustration as they watched the great strides being made in other spheres, whilst they appeared merely to be amassing an enormous pile of facts. When Darwin's book appeared, which provided the opportunity of combining these facts in one comprehensive theory, it was seized upon with uncritical enthusiasm by many. It satisfied the deep expectation of the human mind to find a cosmological theory or law behind a mass of seemingly unrelated facts.

3) THE UNMATHEMATICAL APPROACH

Any serious student of the progress of science would acknowledge the important part which mathematics has played in it, for it has indeed been called the "Queen of the Sciences". It is in this discipline above all where accuracy and exactness are paramount, and errors in formulae or calculations are quickly detected and rectified. Yet it is in this very subject that Darwin admitted his weakness. He had delighted in studying the Euclidean Laws of Geometry, but when it came to the abstract concepts of such subjects as algebra, he was baffled and woefully admitted that: "men thus endowed seem to have an extra sense" [2p46]. His dislike of mathematics is revealed in his reaction to criticism of *Origins* by mathematicians. He wrote:

"It is evidently by Houghton, the geologist, chemist and mathematician. It shows immeasurable conceit and contempt of all who are not mathematicians....there is an article or review on Lamarck and me by W. Hopkins, the mathematician, who, like Houghton despises the reasoning power of all naturalists" [5p153].

On the other hand, Darwin was willing to defend the quite fallacious numerical predictions which Malthus provided in support of his theory on population, for he wrote to Lyell:

"It consoles me that - sneers at Malthus, for that clearly shows, mathematician though he may be, he cannot understand common reason".

One very revealing aspect of this absence of mathematical ability is that it was shared by several of Darwin's close friends. Hooker considered Malthus' arguments were "incontrovertible". Wallace praised him for his "masterly summary of facts and logical induction to conclusions". Clark says that Lyell "was accustomed to express the utmost contempt of mathematical reasoning" [11p75]. Lyell is usually referred to as a 'clear thinker', yet in his first year at Oxford he studied logic for a term, "but made so little progress that he decided to take up mathematics instead" [9p28]. He appears to have been equally poor at this, finishing bottom of his group [7p38].

Huxley similarly showed a dislike of mathematics. In the mid 1860's, Thomson (later Lord Kelvin) showed that age of the solar system was much shorter than geologists and biologists were claiming. Thus the radiant energy of the sun limited its past existence and future duration. The tidal effect on the earth's rotation severely limited its age, whilst its rate of heat loss restricted the period it had been cool enough for life to form [12p49]. Bibby noted:

> "Losing nerve before the weight of the physicists' attack, both the geologists and biologists began to retreat. Darwin himself..... found Thomson an "odious spectre"...".

Huxley, however, counter-attacked in his Presidential Address to the Geological Society in 1869. He challenged Thomson's views, although on what grounds he, as a biologist, was competent to do so is not clear.

He declared:

> "I do not presume to throw the slightest doubt upon the accuracy of any of the calculations made by such distinguished mathematicians....But...this seems to be one of the many cases in which the admitted accuracy of mathematical processes is allowed to throw a wholly inadmissable appearance of authority over the results obtained by them".[12p50]

Huxley's comment is valid in certain circumstances for there is the old saying that "There are lies, damned lies and statistics". This would not apply in this case however. Thomson had based his work upon observed and measured facts which, when analysed by some comparatively straightforward mathematics gave the clear results which he published. Huxley, as a biologist, was in no position to fault either his facts, reasoning or mathematical analysis. Why he so unwarrantably smeared Thomson was that the results contradicted the vast time spans required for Lyell's "Uniformitarian Theory". Yet these time scales, being little more than sheer guesswork, would not have survived a critical examination by any competent geological scientist.

Huxley's criticism of Thompson would surely be more applicable to Malthus, whose mathematical analysis is now accepted as being quite fallacious.

This anti-mathematical outlook was not confined to Darwin and his close friends, and Clark gives two further examples. E.B. Poulton, who said that he looked upon the *Origins* as a Bible, admitted that "Mathematics was as incomprehensible to him as to the master (Moseley at Oxford) he served", whilst H.G. Wells "was completely unable to appreciate physical arguments even in their most elementary forms" [11p76].

[It is incidentally a little surprising that, even though Wells had no grasp of scientific concepts, he should nevertheless proceed to write more than one novel in which he predicts that the scientific era heralded a glorious future for mankind. It was the dropping of the atomic bomb at Hiroshima, however, that made him realize that scientific discoveries could be used also by man, for sinister ends. He wrote his book *Mind at the end of its tether* which is a cry of despair for the situation he could see arising, and for which his humanistic philosophy of life had no solution. This small book, one of his last, however, is usually dismissed as the ramblings of a mind in senile decay.]

This vague, anti-mathematical background behind the theory of evolution is the very atmosphere in which hypothetical speculation is free to run riot, and was a very important factor in the rise and eventual dominance of the theory. Indeed it is the very *exactness* of mathematics, highlighting errors of logical thinking, which Darwin and his collegues would find so disconcerting.

Macbeth has pointed to an abuse of mathematics by some present day evolutionists. He refers to the almost incomprehensible probability calculations used in genetics by Fisher and Sewell Wright. He ironically suggests that a commission be appointed to study the work of these men and that of Ford and Haldane to see if they have added anything whatever to our understanding of nature. He points out that all this was repeated some years ago. Francis Galton (Darwin's cousin) and Karl Pearson applied mathematics to evolution but as they were based upon inaccurate data their results were useless. Macbeth concludes his chapter saying:

"[Sir Julian] Huxley thinks it is different now because the present mathematicians are working 'on a firm basis of fact,' but the day may come when their facts are seen to be erroneous assumptions and their labours go into the wastebasket".[37p53]

A similar criticism could be made of the the profusion of figures in such aspects of evolution as radiometric dating. In this case there is no question of the mathematics being wrong. The objection is with the assumptions on which they are based or the inferences drawn from them.

CHAPTER 19

THE SUBTLE APPEAL

Darwin was well aware that senior naturalists of his day were unlikely to accept his views. It had been proposed at various times and had been rejected by them as unsupported by the factual evidence available in which they were well versed. Darwin deliberately bypassed them, for in the closing pages of his book he appealed to the "young" and "rising naturalists" [p456]. Similarly, he appealed to intelligent "laymen", for in a letter to Asa Gray in America dated 21 December 1859, he wrote:

"I have made up my mind to be well abused; but I think it of importance that my notions should be read by intelligent men accustomed to scientific argument, though *not* [emphasis his] naturalists. It may seem absurd, but I think such men will drag after them those naturalists who have too firmly fixed in their heads that a species is an entity" [3p245].

Thus Darwin was aware that his views would be most eagerly accepted by the young, the ambitious, and the layman. Why should these three groups be most susceptible to his views?

A) THE YOUNG

It is generally agreed that new ideas are more readily accepted by the young than by older people who would be more set in their ways. First they would lack the years of mature wisdom which would enable them to judge adequately the strengths and weaknesses of many arguments used. It is for this reason that such theories as Marxism (which nowhere has ever provided for the people of any nation the benefits it always promises) find its most fanatical adherents among the student population of the universities of the world.

Secondly, there is a natural desire in all young people to establish their independence from the generation of their parents. A useful means to this end is the adoption of new ideas in order to drive a wedge between themselves and the older generation. 'New' is considered to be 'best' and the older generation is dismissed as out-of-date or 'square'.

Darwin's appeal certainly succeeded with the young scientists of his day and not in this country alone. In 1864, Hugh Falconer wrote to Darwin, and referring to a French professor, said:

"He told me in despair that he could not get his pupils to listen to anything from him except a la Darwin. He poor man could not comprehend it, and was still unconvinced, but that all young Frenchmen would hear or believe nothing else" [5p257].

B) THE AMBITIOUS

Darwin's appeal to the "young and *rising* naturalists" is a subtle hint to those who were ambitious in making their mark in the world to use his theory in order to achieve their ends. Indeed, he followed this by suggesting that those who *do* accept the theory should "express their conviction " so that the "prejudice" should be "removed".

Certainly, his book had great impact amongst many young scientists of his day who were later to become leading figures in their respective fields. Himmelfarb gave some examples of this. A. Geikie, the geologist, referred to the book as a "new revelation". N. Shaler, the American geologist, debated the theory in secret as if it were a heresy. August Weismann, who was later to disprove Darwin's (and Lamarck's) idea of the inheritance of acquired characteristics, said:

> "Darwin's book fell like a bolt from the blue; it was eagerly devoured, and while it excited in the minds of the younger students delight and enthusiasm, it aroused among the older naturalists anything from cool aversion to violent opposition".

C) THE LAYMAN

As we have related, the study of Geology was a very popular hobby amongst both men and women of Darwin's era. This was largely the result of the interest aroused by Lyell's book *Principles of Geology*. There was therefore a considerable body of amateurs who would be interested in Darwin's theory, particularly as it was in harmony with Lyell's Uniformitarian Theory of the earth's strata and fossils. It thus became the chief topic of conversation at many of the fashionable dinner parties.

Interest was not confined to the upper strata of society alone for Huxley gave lectures at a Working Men's Institute entitled *The Relation of Man to the rest of the Animal Kingdom*. At that time there was not a single piece of fossil evidence linking man with the apes, yet Huxley boasted to his wife:

> "By next Friday evening they will all be convinced that they are monkeys...Said lecture let me inform you, was very good. Lyell came and was astonished at the magnitude and attentiveness of the audience"[13 p276].

Darwin also appealed to men of 'intelligence'. It may be asked which of his readers, having read his intricate book, would not consider himself intelligent? But intelligence alone is not sufficient to determine whether or not a theory is true. To do so requires access to a considerable body of facts which few would enjoy, and most would therefore be incapable of refuting any scientific theory presented. However, whilst the ordinary laymen are unlikely to have a major effect upon scientific theories, they do create the climate of opinion which would allow certain concepts to spread more swiftly.

More important than this, however, was the opportunity which the theory presented to the layman who was unlikely to understand the latest scientific developments, particularly if he shared Darwin's lack of mathematical ability. Such a person, once he had grasped the principles of this basically simple theory of evolution, could now think of himself as 'scientifically minded'. Clark put this very well, for he wrote:

"One outstanding achievement of Darwin lay in the fact that he broke the spell of physical science. Henceforth "science" became the monopoly of all who chose to apply the word to their own studies. Marx, deeply stirred by Darwin's writings, could appeal to the working classes to rebel in the name of a "science" of history. Capitalists and politicians could claim that "science" was on their side. Henceforth vague and unsupported conjectures about the development of language, of race, of cultures, of religion, or of social customs, could all be called "scientific." Darwin showed how the journalist, the popular writer, the political propagandist - all, in fact, who resented exact or disciplined thought because they could not be bothered to master it, could have their revenge for the inferiority of feelings that it had occasioned them. Nor were the pseudo-scientists slow to follow Darwin's lead.

This violent revolt from exact science is illustrated best of all, perhaps, in the writings of Herbert Spencer, who poured forth volume after volume of the wildest speculations in an attempt to provide a "scientific" philosophy of life. Needless to say, the exact scientists of the day were astounded and revolted by his activities." [11 p76-7].

We have now examined the reasons why the three groups of the young, the ambitious and the laymen should most readily accept this revolutionary new scientific theory. There is one factor which is common to the attitude of all three groups, which is their basic motivation or human weakness, depending on how you view it. This factor is fully covered in one word.

PRIDE.

Darwin's subtle appeal to the pride of his readers is most clearly seen in one paragraph of his conclusions which occurs just seven pages from the end of his book. I have already quoted from this section, but I consider it so important that I give it in full:

"Although I am fully convinced of the truth of the views given in this volume under the form of an abstract, I by no means expect to convince experienced naturalists whose minds are stocked with a multitude of facts all viewed, during a long course of years, from a point of view directly opposite to mine. It is so easy to hide our ignorance under such expressions as the 'plan of creation,' 'unity of design,' etc;, and to think that we give an explanation when we only re-state a fact. Anyone whose disposition leads him to attach more weight to unexplained difficulties than to the explanation of a certain number of facts will certainly reject the theory. A few naturalists, endowed with much flexibility of mind, and who have already begun to doubt the immutability of species, may be influenced by this volume; but I look with confidence to the future, - to young and rising naturalists, who will be able to view both sides of the question with impartiality. Whoever is led to believe that species are mutable will do good service by conscientiously expressing his conviction; for thus only can the load of prejudice by which this subject is overwhelmed be removed." [p456].

This passage deserves close examination both to demonstrate that it is a thinly veiled provocation of the pride in his readers and to expose the devious way in which he does this.

First, he says that experienced naturalists will not agree to his views simply because they will not look at them from his point of view. However, most of them rejected evolution as they considered it to be against the *facts*, which is quite a different matter.

He then accuses them of plastering over the gaps in their ignorance by referring to the "plan of creation" as being no explanation, only a restatement of fact. But which are the more important: acknowledged facts or the hypothetical 'explanation' which Darwin wished to elevate above the facts? On this point, the creation of species is a perfectly adequate 'explanation' and is more in accord with the facts than evolution. Darwin attempted to soften his accusation by using 'our' and 'we', but it is obvious he does not include himself in the group he describes.

Following this he flatters those who accept that species can develop (that is, his supporters) as being amongst the elite "few...endowed with much flexibility of mind." In this way he was inviting them to sneer at their opponents as narrow-minded bigots, a tactic still frequently employed by evolutionists today.

His call to the "young and rising" generation of naturalists to express their "conviction" has already been noted, but he then goes on to say that only they will be able to judge "impartially" on the issue.

He finally concluded his appeal by saying that the opposition which they will meet is simply due to "prejudice", implying of course that those who agreed with him were quite free of any such failing!

Careful examination of this passage demonstrates how Darwin deflected attention from the factual weaknesses of his case by confidently claiming he was right and ridiculing the "prejudice" of his critics. In the reading of this paragraph, one is reminded of the story of the politician who, in preparing a speech, added at one point the marginal note: *"Argument weak - shout louder"*!

Summary

This examination may be concluded by referring to this use of the term "prejudice" which Darwin used to describe his critics, but which could also apply to those who accepted what was only a hypothesis, as Darwin generally conceded. I would question whether *anyone* can be absolutely certain that he has ever made a judgment on an important subject completely free of any prejudice whatsoever. Indeed, the theory of evolution, which contains the possibility that there is no God, cannot be judged independently of one's philosophy of life. In fact I would go further and suggest that a person's philosophy *pre-determines* his view of the theory of evolution. This subject is of such importance that we will return to it in a later section.

CHAPTER 20

THE RESULTS

The results of evolutionary principles has obviously been very profound and warrants a study on this aspect alone. In this section however we will examine its influence in the three areas of science, religion and politics where its effect has been far from beneficial.

1. SCIENTIFIC INVESTIGATIONS

Occasionally one reads that evolution provides the essential framework for scientific investigation, and without it, the work of many scientists would have no relevance to the established body of scientific knowledge. Such claims however are unjustified, for evolution only provides a theory of how organisms came to exist as they are. Whether true or false, it is irrelevant to the investigation of how such organisms actually operate.

Another claim is that evolution has stimulated an enormous amount of research into many fields of science. Invaluable facts have come to light which have been in many cases of great practical significance. Again the same criticism can be made, for these are the results of the study of living forms. It is difficult to see how a knowledge of the evolutionary path, along which an animal developed, can be of any practical significance in changing its metabolism or utilizing it in any way. We have to study organisms as we find them today, and work from that point onwards.

Far from assisting in the growth of knowledge, it could be claimed that the theory has actually *hindered* the progress of science. Scientists who support evolution have spent many hours of research time and facilities (and on occasions expensive equipment) simply to obtain much sought for 'links' or 'relationships' between various animals in an attempt to prove descent. Often, the connection established is far from convincing and frequently conflicts with other evidence. Several examples are given in Patterson's book *Evolution*, i.e. the various relationships between man and the primates [22p140].

One of the most well documented cases of fruitless investigation prompted by the theory of evolution is that of the 'Biogenetic Law of Recapitulation' (i.e. an embryo recapitulates the stages of its primordial ancestors). This theory was aggressively propagated by Professor Haeckel of Jena University. Much effort was expended by scientists to discover by this means the genealogy of many animals, whilst Haeckel bitterly opposed the fundamental and important research of Professor His which he dismissed as being "irrelevant".

Today however it is acknowledged that the whole theory is quite erroneous. Sir Gavin de Beer, himself an evolutionist, has stated:

"Seldom has an assertion like that of Haeckel's 'Theory of Recapitulation', facile, tidy, and plausible, widely accepted without critical examination, done so much harm to science" [54p159].

One cannot but wonder how much effort, time and money could have been put to better use in many laboratories had scientists entered their field of research without similar evolutionary presuppositions.

2. THE ANTI-CHRISTIAN

Prior to the publication of *Origins*, there was a general acceptance of the existence of God and that He was the wise and benevolent creator of the universe which scientists and naturalists alike were exploring. This had found its expression in Paley's work, in which he logically deduced the existence of a creator from the beautiful design of the creatures he made. Scientific papers not infrequently referred to the Almighty with no embarrassment to the writer or reader.

This is not to say of course that there were no anti-Christian feelings, but these generally expressed themselves not so much by direct attack but by the proclamation of an alternative philosophy, such as the evolutionary writings of Darwin's many predecessors. Such forces were more latent than effective, however, because of the overwhelming acceptance of Christian beliefs by the large majority of people, even though in retrospect much of it appears to have been little more than "lip service".

It was Darwin's book, however, which was to change this climate of opinion, for it became the rallying point from which more determined attacks, now carried forward under the banner of "science", could be mounted against the Christian church. Darwin, although careful never to refer to the religious implications of his views, nevertheless presented the unbelieving man with an ideal means by which he could "free" himself from the all-seeing, righteous God of Christianity. It is in response to this call to the deep-seated desire of men that I would attribute the rapid acceptance of evolution in Darwin's time, and its increasingly strident proclamation in the mass media still today.

3. THE POLITICAL JUSTIFICATION

One of the most intriguing aspects of the outworking of the evolution theory is the way in which it was seized upon by both the extreme right and left wings of the political spectrum, as a justification of their actions.

A) RIGHT WING

Herbert Spencer, as we have seen was an extremist in allowing the "survival of the fittest" to take its barbarous course in its outworking in human society. Clark gives other examples of tycoons who justified their ruthless exploitation and sharp dealing by voicing the well worn catch phrases of the evolutionist's dogma of faith. Andrew Carnegie was deeply perturbed by the un-Christian ways of big business, but having read the works of Spencer, his concern was overcome and he claimed: "While the law may be hard for the individual, it is best for the race". Similarly, J.D. Rockefeller, who amassed a fortune by scandalous methods, said in an address to a Sunday School gathering:

"The growth of large businesses is merely the survival of the fittest.... It is merely the working out of a law of nature and a law of God".

It is interesting to note that the theory was much more enthusiastically received in Germany than it was in this country, being accepted as "proven fact" much earlier. In 1877, when attempts were made upon the life of the German Emperor, the theory of evolution was blamed for spreading discontent among the workers. Ernst Haeckel immediately defended the theory, claiming that it was right that only a "small and chosen minority" of the "fit" should survive!

The evolutionary basis of the "superman" theories of Nietzsche and the use which Hitler made of his theories in his "Master Race" plan are well known. The frightening outcome of taking the evolutionary theory to its logical conclusion by eliminating the "weaker" races is displayed in the horror of the concentration camps of Belsen, Ausschwitz, etc. What is of interest is that, whilst historians carefully document the atrocities of Hitler's regime, not one will highlight the fact that it was based upon an evolutionary philosophy. The theory is indeed sacrosanct and guarded from criticism from any source.

B) LEFT WING

Karl Marx's claim that "Religion is the opiate of the people" shows that godless evolution would be virtually the only philosophy of the natural sciences which he could possibly combine with his political ideas.

Marx rejoiced that the death blow had been dealt to teleology - the evidence for God in the natural world. Marx was impressed with Darwin's theory and studied it carefully. It is sometimes stated that he offered to dedicate the English translation of Das Kapital to Darwin - an offer which was refused. This has since however been found to be wrong as we will show later.

Lenin similarly "studied Huxley's views on agnosticism, materialism, etc., with interest and care" [12p170].

The importance of evolution in Communist philosophy and the fanatical zeal with which it is held and propogated is too well known to require further comment. Yet strangely there are two fundamental aspects where evolution contradicts the Communist philosophy.

A) The first objection is that if evolution has taken place and is still operating, then surely those in positions of power, i.e. the capitalists, could be considered as the "fit" who have "survived" - as they themselves claim.

If this were so, then the present situation should be allowed to persist, so that evolution is left to continue on its course free of interference by the efforts of men.

B) The second is that evolution (as originally proposed) claimed that the large differences between species are the result of numerous small modifications which take place over vast periods of time. Communism on the other hand tries to bring about social change by a violent action. Attempts were made to overcome this problem by talking about a gradual increase of small changes suddenly resulting in the appearance of a large change, but this was not accepted as it was contrary to the way in which evolution was supposed to operate.

This view of evolution slowly making small changes in species during millions of years is the one proposed by Darwin and accepted ever since his time. In the last few years however there has arisen an alternative view of how evolution operates which has allowed the Marxist philosophy of political change and the theory of evolution to come very much closer together.

I have already quoted the denunciation of Lyell's (gradualist) Uniformitarian Theory by the evolutionary geologist, S.J. Gould of Harvard University. Gould points out that the strata show clear evidence of catastrophes. Similarly his colleague, Professor Eldrege of the American Natural History Museum has admitted that the geological column, far from showing a steady evolutionary progress, has very large gaps between major groups. The existence of these large unbridged gaps between major groups and species is, of course, one which creationists have been making for many years.

Why should prominent evolutionists such as Gould and Eldrege now make admissions which appear to be so damaging to the theory?

Their whole purpose is to propagate a variation of the Theory of Evolution called Punctuated Equilibrium. This proposes that there were long tranquil periods of stability which were 'punctuated' by periods of rapid change. Such a view of evolution is exactly paralleled by the Marxist philosophy that political changes are brought about by numerous small pressures and influences which finally culminate in a rapid change of political power.

In an article [66] they say that Darwin's 'gradualist' theory was very much a product of the prevailing philosophy of his day, and that "Alternate conceptions of change have respectable pedigrees of philosophy" and refer to the ideas of Hegel and Engel "which have become the official 'state philosophy' of many socialist nations" [p145]. In pointing this out they demonstrate so clearly the flexibility which the theory possesses in adapting itself to any of the prevailing materialist world views!

What is perhaps very significant is that Gould is a Marxist and the article mentions "It may also not be irrrelevant to our personal preferences that one of us learnt his Marxism, literally at his daddy's knee"[p146]. Such a fact is surely far from irrelevant, for I would suggest that it forms the whole basis of his approach to the subject.

Clearly what is happening is that Marxist biologists are pointing to the fossil record as proof that such rapid changes have occurred in the past and by inference will occur in the future. These views are now receiving an increasing amount of publicity, such as the BBC TV programme *Did Darwin get it wrong?* A not dissimilar approach can be seen in the frequent use of 'cladograms' now appearing in the new exhibitions at the British Natural History Museum. This whole subject and the claim by an evolutionary geologist that the Museum has been influenced by Marxist principles will, however, be considered in a later chapter.

The points which I have made above will doubtless be completely ignored by the Left Wing theorists, for evolution is the only philosophy whereby they are enabled to eliminate the concept of God. This recent development of Punctuated Equilibrium is simply changing the theory so that it can be used more effectively in propogating revolutionary ideologies.

CHAPTER 21

DARWIN — THE MAN

What sort of man was it who produced such a work that was to dominate the thinking of generations to come. Was he an honest man setting out his ideas to the best of his ability, or a subversive intriguer intent upon producing a new theory which would deceive millions? I would suggest that (to begin with at least) he was neither.

Thus far, I have clearly been extremely critical of Darwin's methods and motivation. However, I must make it clear that with regard to his general personality and relationships with his family and friends, by all accounts he was a very warm and pleasant character, noted for his gentle kindness. He was an excellent husband to his wife, Emma, and deeply loved his children, all of whom greatly respected him and responded to his sincere affection. His grand-daughter, Gwen Raverat, paints a most affectionate portrait of his children in their later life in her delightfully charming book *Period Piece* [43]. Similarly in his dealings with his scientific colleagues, he was generous in his opinions of many of those who opposed him.

All these characteristics are usually well featured by his biographers and I will therefore not dwell on them here at length. As I have already said, the purpose of this work is to highlight those aspects which receive little attention, particularly where they play an important part in the formulation of his ideas.

A) SCIENTIFIC DISHONESTY

I have already referred to Darwin's devious arguments, his carefully selected use of "evidence", etc., that there is no need to repeat the charges here. I would, however, return to one matter where he deliberately side stepped the evidence which completely refuted the whole basis of "his" theory.

In order to obtain evidence for his work, Darwin took up pigeon breeding and frequently had dealings with experienced breeders of a variety of animals. He must therefore have been fully aware of the fact that there are limits to which any animal can be bred by selection of characteristics. How did he answer this fundamental objection? At the end of the first chapter of *Origins*, he casually remarks that:

> "Some authors have maintained that the amount of variation in our domestic production is soon reached and *can never afterwards be exceeded*".

He answers this by continuing:

> "It would be somewhat rash to assert that the limit has been

attained in any one case; for almost all our animals and plants have been improved in many ways within a recent period; and this implies variation. It would be equally rash to assert that characters now increased to their utmost limit could not, after remaining fixed for many centuries, again vary under new conditons of life".

He then spends the rest of the paragraph dealing with *variations*, and completely fails to answer this most serious objection to the theory. He merely claims that species do vary, and such limits as are now known *might* be enlarged in a few centuries' time "under new conditons of life"! Surely, virtually to ignore this very unwelcome, but in his day, well established fact that species vary only within limits, is to be guilty of a lack of scientific integrity. This feature would surely prejudice Darwin's claim to be an objective scientist who carefully collected his evidence and from this alone to postulate theories "on true Baconian principles". [In chapter 26 we will be considering some recent developments regarding varieties and genetics].

It is interesting that Charles Lyell also recognized the problem which the simple variability of species presented. In a letter he wrote in May 1860, he said:

"With *limited* [emphasis his] variability, *which is an arbitrary assumption after all,* [emphasis mine] we can explain nothing" [8p333].

Thus the founders of evolution, recognizing the serious problem which limited variability formed, and about which the public would be generally ignorant, called it an "assumption" and shamelessly glossed over it. With this barrier removed, the floodgates of human imagination were opened and hypothetical genealogies could now be constructed by the most uninformed amateur evolutionist of the day.

Furthermore it must be asked, what sort of 'explanation' were they looking for? All the available evidence indicated that species (or major groups) had been specially created but were able to vary within certain limits. This was a perfectly satisfactory 'explanation' and was the accepted opinion of most scientists of the day. Yet Lyell sought another 'explanation' which I would suggest is merely *any* mechanistic theory which would account for the existence of life *without requiring the presence of an all powerful creator God.*

B) HEALTH

Throughout the voyage of the Beagle, Darwin was a healthy extrovert, stronger and fitter than most of his companions. Shortly after his return, however, he had an onset of illness which was to plague him for the rest of his life. The symptoms were sickness, headaches, sleeplessness, faintness, twitching muscles, spots before the eyes, and others, all of them being due to a nervous disorder. The attacks were often severe and prolonged, preventing Darwin from working, accepting engagements and inviting or visiting friends.

There has been much speculation on the cause of his complaint, but little evidence. The cause usually mentioned is that he caught Chagas's disease when he was bitten by an insect in South America. It is claimed that this disease would explain Darwin's symptoms precisely. I am unable to comment on whether such a diagnosis is correct or not. I can only voice my suspicion that it would be in the interest of the propagandist of the theory of evolution to protect Darwin's reputation from any charge that he was emotionally unstable for this would cast a blight upon the image of him as a "clear thinking scientist". To this end, Darwin's life would be minutely examined in order to provide some purely medical cause for the severe symptoms he suffered. The note in his diary that during his travels he was bitten by an insect which was later said to transmit Chagas' disease amply fulfilled the role required. All of Darwin's symptoms however are capable of being diagnosed as due entirely to emotional stress.

Darwin resented the accusation that he was a hypochondriac, yet there is some justification for it. Indeed his family history has a record of illness. His grandfather, Erasmus, stuttered badly and was rather odd, whilst Erasmus's son suffered from a mental illness and committed suicide. Darwin's brother (also named Erasmus), suffered from melancholia, whilst Darwin's own children were hypochondriacs to varying degrees.

Clark, in his analysis of the cause of Darwin's state is quite categoric, for he says:

"The answer to the riddle is not far to seek. Darwin's trouble almost certainly lay in the suppression of his religious needs. His life was one long attempt to escape from Paley, to escape from the Church, to escape from God. It is this that explains so much that would otherwise be incongruous in his life and character" [11p85].

He proceeds to show how Darwin consistently refused to acknowledge any element of design in the world, either by dismissing it or bypassing it with the joke that if everything was designed, this must include every single detail of events, including the shape of his nose!

I do not think, however, that this gives a completely satisfactory explanation of the severe symptoms of mental stress which Darwin was undergoing. Darwin seems to have been fairly unperturbed by the gradual loss of his Christian faith. Similarly, multitudes turn away from their faith and yet suffer no adverse physical symptoms. The following alternative reason however may explain the symptoms more satisfactorily.

That there is a connection between Darwin's theory of evolution and his ill health is suggested when we consider their timing. As we have seen, Darwin's first thoughts on evolution are given in a notebook started about July 1837. The very first signs of his illness are referred in a letter written in the autumn of the same year, for he mentions:

"I have not been very well of late, with an uncomfortable palpitation of the heart, and my doctors urge me *strongly* [emphasis his] to knock off work....".

Thus, within a month or two of setting out the whole structure of the theory, he has an onset of symptoms of stress. Similarly, at the other end of his life, when he had finished writing "The Descent of Man" in 1871, his health improved considerably.

From this it would seem that his illness was linked to his writing on the "controversial" subject of evolution. I would suggest that the root cause of Darwin's illness was the stress generated in him when he was writing about a theory *which he knew was basically false*. All his symptoms are those of a man who is under prolonged emotional stress — stress due to his continual mental acrobatics as he sought to "wriggle" (a word he used to describe his arguments) around a whole series of facts against his theory.

What inner promptings could have driven him to continue along this course, regardless of its effect upon his mental and physical state? Not a desire to avoid God, but an inordinate hunger for fame and recognition – a factor which we will consider in some depth later.

C) CHRISTIANITY

As we have mentioned, Darwin studied for holy orders and was greatly impressed by Paley's arguments for the existence of God in the evidence of design in the workings of nature. On the *Beagle* he accepted the authority of the Bible and was laughed at by his fellow officers when he quoted it to settle a dispute. He also referred in his notes to "centres of creation", in the normally accepted sense. Thus, as he admits, he was perfectly orthodox in his belief until his return to England, from which date his belief began to decline.

In 1876 Darwin wrote:

"Thus disbelief crept over me at a very slow rate, but was at last complete. The rate was so slow that I felt no distress" [2p309].

He finally claimed he was an agnostic but never an atheist. He was visited by Dr. Aveling (Karl Marx's brother-in-law), who later wrote a pamphlet for the Free Thought Publishing Company in 1886, in which he claimed that Darwin's form of agnosticism was virtually atheism. This, however, was denied by Francis Darwin, who edited his father's papers [2p317].

In order to set the record straight for the many millions of viewers of the television series *The Voyage of the Beagle*, one incident which did *not* turn Darwin against the Christian faith was the death in 1851 of his daughter, Annie, at the age of ten years. In the series he is portrayed as becoming very bitter about the death of his greatly loved child. Yet in an account he wrote of her death a few days later [2p132], there is sorrow and a very affectionate remembrance, but no trace whatsoever of bitterness towards a God who could allow it to happen.

A return to faith ?

It is at this point that I will refer to a report that in his last years Darwin became a convinced Christian. The fullest account appears in pamphlet no. 80 of the Creation Science Movement. A Lady Hope is said to have visited Darwin shortly before he died and gave an account of their meeting at the evangelist, D.L. Moody's educational establishment at Northfield, Boston. The subject however is controversial, far from certain and not ultimately of great significance to the progress of the theory. It does however interest many Christians and I recount the incident and some recent discoveries in Appendix I.

D) VANITY

Darwin's attitude to the public acclaim which he eventually received was largely one of retiring modesty. His chronic illness invariably prevented him from attending various meetings to receive honours, medals, etc., which were showered upon him, and he may well have used this excuse at times to avoid appearing too much in the limelight. This would suggest that he had a streak of shyness in not wishing to be the centre of attention in very large gatherings, which points to a commendable modesty.

This would not prevent him, however, from possessing a strong desire to be recognized as a "famous" person. One example which I have already mentioned is his continual reference to evolution as "my theory" throughout his *Origins*, even though every single one of his ideas had been proposed by others before him.

A further seemingly small example of Darwin's vanity is his recollection in his autobiography of a remark by a famous politician of the day who, in reference to the young eighteen year old Darwin, said: "There is something in that young man that interests me". As Himmelfarb comments: "It is pathetic to think of him cherishing the memory of this one compliment" and recording it fifty years later.

Even the reason he gives for writing his autobiography is rather strange. He states that "the attempt [to write it] would amuse me, and might possibly interest my children or their children". He knew he was an internationally famous person, yet he made no suggestion that his autobiography was strictly for private family use. *Therefore, he must have known that it would be published worldwide*, as indeed it was, being placed at the beginning of Volume I of the first collection of his letters [2]. Thus, whilst modestly claiming to have only private objectives, he nevertheless was well aware that his writings would receive world-wide publicity.

In his autobiography, Darwin admitted that as a child, he was "much given to inventing falsehoods, and this was always done for the sake of causing excitement". He described how, as a child, he boasted to a friend that he could change the colours of flowers by watering with coloured fluids, which he admitted was "a monstrous fable, and had never been tried by me". He also hid some fruit and then claimed he had "discovered a hoard of hidden fruit".

One would not of course wish to judge a man on such small misdemeanours of his youth. However, when taken in conjunction with other factors, it seems likely that his desire to be the centre of attention, *irrespective of the means*, may have been so deeply rooted in him that he never really outgrew it in adult life. It is difficult to assess a person's real character from his writings, but I strongly suspect that beneath a seemingly retiring personality lay a streak of vanity which was deeper than for many men. This occasionally surfaced in some of his writings, such as his comment in his autobiography that, whilst on the voyage of the *Beagle*, one of the reasons he worked hard was that he was "ambitious to take a fair place among scientific men", but adds humbly "- whether more ambitious or less so than most of my fellow-workers, I can form no opinion". I would contend that Darwin's ambition for fame is more than sufficient to explain why he so consistently ignored or evaded the great body of facts against evolution and made the most of what evidence he could collect in writing his *Origin of Species*.

Was there any particular influence which could have resulted in Darwin having a great desire for fame? Those who look for psychological factors would not have to search far.

E) DARWIN'S CHILDHOOD

Darwin, when writing about his father, always referred to him with warm appreciation and respect, only once mentioning that "he was a little unjust to me when young".

However, the real atmosphere in the Darwin household was more accurately described by Emma Darwin, who gave an account to their son, Francis, shortly after Darwin's death. The original manuscript is held at Cambridge, but this account was omitted when her letters were published in *Emma Darwin – Letters*

[1 p377]. She often visited the house as a child and referred to Darwin's father as tyrannical and unsympathetic:

"Everything in the household had to run in the master's grooves, so that the inmates had not the sense of being free to do just what they liked", and she was convinced that "Dr. Darwin did not like him or understand or sympathise with him as a boy" [1 p7].

An even more revealing comment on the atmosphere which existed in Darwin's family was given by Irvine who mentioned that his father "invariably concluded his day's work with a two hour monologue to his awe-stricken children"![16 p34]. Such a stressful atmosphere may well explain why Darwin's brother, Erasmus, was emotionally crippled for much of his adult life, suffering from chronic listlessness and depression

When, to all this, is added the fact that Darwin remembered nothing about his mother, who died when he was eight, it is not difficult to imagine him developing a strong subconscious craving for the love and affection which all children need, particularly in their early years. Those who are deprived of such love when children often indulge in an escalating series of acts, each more outrageous than the last, simply to draw attention to themselves in an attempt to fill the void left by inadequate affection. When, furthermore, they are subjected to an over harsh discipline when young, as Darwin was, they often "go off the rails" in later years when they obtain a degree of freedom. This is precisely what Darwin did when he mixed with disreputable types on going to University.

When such basic needs of a child go unfulfilled, they invariably affect its personality throughout the remainder of its life. Irvine admitted that "Charles was left with a morbid craving for affection". I would suggest that this would explain Darwin's latent but strong desire for fame, which was to overcome the more generous and retiring elements in his nature with which it was bound to conflict.

In conclusion it would appear to me that Darwin was guilty of deliberately deceiving the world by postulating a theory which he knew to be false and his action wreaked its own revenge upon his physique in this life.

However, whilst not wishing to minimise the seriousness of the charge I am making against Darwin, any judgment we may pass should be tinged with understanding. With Darwin's childhood in mind, can any one of us be certain that we would have had any greater strength to resist such powerful internal forces? "There but ..."

Having studied Darwin's life in some detail, I see not a man patiently collecting facts in support of a scientific theory, nor a scientist boldly facing the experts with his startling new theory, but I see a small child, desperately crying out for the love, security and recognition which had been denied him.

Perhaps the greater blame should be charged against those who, playing upon Darwin's weaknesses, used him simply as a tool for propagating their godless views upon a public whom they had already prepared to receive them.

SECTION IV

THE SECRET AIM - ACHIEVED

CHAPTER 22

LYELL'S REAL MOTIVE

As we have shown, Darwin, throughout the voyage of the *Beagle* and after his return, was a creationist, and did *not* begin to doubt the fixity of species whilst he was on the Galapagos Islands. What could have happened that he should have started to have set out in his notebooks, ten months after his return, what was eventually a complete outline of the theory of evolution that was little different from his book published twenty two years later? The answer, I am sure, becomes clear when we study the life of Charles Lyell and the very great influence he had upon Darwin.

Lyell, when writing his *Principles of Geology* was very cautious how he referred to the question of "the fixity of species". Whilst saying that species were fixed, he nevertheless used very circuitous arguments and left it very open for it to be interpreted that species *could* have changed. Himmelfarb states that Spencer was converted to evolutionism by his reading of Lyell's *Principles*. Similarly, Darwin, shortly after his return, was surprised to find "Lyell much more tolerant of evolution than his *Principles of Geology* indicated" [16p43].

Irvine notes:

"The second volume of Lyell's "Principles" was really the "Origin of Species" without Darwinism, or at least explicit Darwinism. In almost the same sequence, Lyell took up the problems of the "Origin"......and did everything but solve them." [16p58]

Why Lyell should have adopted seemingly creationist views becomes clear from his personal letters of that time. As early as 1826, before he had published his book, he was already a convinced evolutionist, having read Lamarck on the subject. Lyell knew, however, that to propose evolution as well as his Uniformitarian theory would have roused the religious leaders and scientists of the day, for they would immediately realize that this was an attack upon their creationist views. He therefore appeared to take their side, whilst so wording his arguments that an alternative interpretation could later be put upon them. His deliberate aim of not offending the deeply held religious views of others, in order to gain their confidence, was but one part of his long term scheme.

Much to Darwin's dismay, Lyell never committed himself fully to accepting the theory of evolution. Nevertheless, he advised Darwin how he could make his case stronger and he was said to be "a great mover of men" who was preparing "a bodyguard of experts to shout down the howl of execration" which was sure to greet the *Origin*. As we have seen, it was his efforts which successfully induced the publisher to accept Darwin's book.

Just how devious Lyell was in covering his real intentions can be seen from a letter he wrote to Darwin in 1863 for he said:

"But you ought to be satisfied, as I shall bring hundreds towards you, who if I treated the matter more dogmatically, would have rebelled"[8p363].

With all this activity, he clearly wished the theory to have as much publicity, support and effect as possible. Yet he never publicly accepted a full commitment to the theory. Could this have been due to purely scientific caution?

I suggest that the simplest explanation of his conduct is that he was well aware of both the scientific weakness of the theory and the storm it would generate. Should the whole case collapse, he had no wish to be buried with Darwin beneath the rubble. In his writings, therefore, he paid just sufficient lip service to the creationist view to be still accepted as one of them, should the worst come to the worst. Thus he would live to fight another day in order to quietly propagate his views.

Intrigue against the "Mosaic account"

Lyell's subtle yet effective attacks upon the accepted view of the day that all the strata were laid down at the time of the Noachic deluge is seen in various passages of his *Principals of Geology*. For example, he spends the first four chapters giving a review of the history of geology, and those who hold to the "Diluvial Theory" are gently ridiculed and described as "prejudiced". For example, in dealing with Scilla (1670) he says :

"This work proves the continued ascendancy of dogmas often refuted....Like many eminent naturalists of his day, Scilla seems to give way to the popular persuasion, that all fossil shells were the effects and proof of the Mosaic deluge. It may be doubted whether he was perfectly sincere.... "[20p37].

This passage is immediately followed by one in which he deals with the Diluvial Theory itself. He says:

"The Theologians who now entered the field in Italy, Germany, France and England were innumerable; and henceforward, they who refused to subscribe to the position that all marine

organic remains were proofs of the Mosaic deluge, were exposed
to the imputation of disbelieving the whole of the sacred writings..
...Never did a theoretical fallacy, in any branch of science,
interfere more seriously with accurate observation and the sytem-
atic classification of facts....[The diluvialists] saw the phenomena
only, as they desired to see them, sometimes misrepresenting
facts, and at other times deducing false conclusions from correct
data."

"....It may be well, therefore, to forewarn the reader, that in
tracing the history of geology... he must expect to be occupied with
accounts of the retardation, as well as of the advance, of the
science....In short, a sketch of the progress of geology is the history
of a constant and violent struggle of new opinions against doctrines
sanctioned by the implicit faith of many generations, and supposed
to rest on scriptural authority. The enquiry, therefore, although
highly interesting to one who studies the philosophy of the human
mind, is too often barren of instruction to him who searches for
truth in physical science."[20p38]

In this important passage we can see how cleverly Lyell equates
all those who hold to the Diluvial theory as "interfering" in the
collection of facts, "misrepresenting" those facts they did obtain,
and holding to "doctrines" which are simply due to the "faith of
many generations" that are "supposed to rest upon scriptural
authority". All this is due to the "philosophy of the human mind"
and is "barren of truth". One can imagine the effect which such
passages would have upon those who were fearful of being associated with an ancient theory which was now so clearly disproved in
a "scientific" work of three volumes by such an eminent authority
(albeit an amateur) as Lyell on the subject. We have already seen
however just how unscientific Lyells researches were, as well as his
deliberate forcing of them to fit his Uniformitarian theory.

Lyell reveals his true aims

We will now refer to a letter written by Lyell which is crucial to
our understanding of the rise of evolution, for it reveals his basic
attitudes and underlying motivation.

The letter was one he wrote in 1830 to a geologist friend,
Poulette Scrope, who was about to review the first volume of his
Principles in the *Quarterly Review*. His letter is full of points which
he suggests Scrope should make or avoid in his review. Regarding
the reactions of the Church establishment, he says:

"I am sure you may get into Q.R. what will free the science
from Moses, for if treated seriously, the party are quite prepared
for it. A bishop, Buckland ascertained (we suppose Sumner),
gave Ure a dressing in the *British Critic and Theological Review*!
They see at last the mischief and scandal brought on them by
Mosaic systems. "[7p270]

His reference to "the Mosaic systems" is a typically oblique reference to the accounts of Creation and the Flood, as given in Genesis. It is at the end of this letter that we come to the passage of particular interest. In this he says:

"If we don't irritate, which I fear that we may (though mere history), we shall carry all with us. If you don't triumph over them, but compliment the liberality and candour of the present age, the bishops and enlightened saints will join us in despising both the ancient and modern physico-theologians. It is just the time to strike, so rejoice that, sinner as you are, the Q.R. is open to you. If I have said more than some will like, yet I give you my word that full half of my history and comments was cut out, and even many facts; because either I, or Stokes, or Broderip, felt that it was anticipating twenty or thirty years of the march of honest feeling to declare it undisguisedly. Nor did I dare come down to modern offenders. They themselves will be ashamed of seeing how they will look by-and-by in the page of history, if they ever get into it, which I doubt..."

"P.S. ...I conceived the idea five or six years ago, that if ever the Mosaic geology could be set down without giving offence, it would be in an historical sketch, and you must abstract mine, in order to have as little to say as possible yourself. Let them feel it, and point the moral" [7p271].

A careful examination of the passage is particularly illuminating.

Firstly, we see how disdainful he is of the whole of the church authorities, not just those who will oppose him, but those who fall for his flattery also.

Secondly, he is well aware that his main opposition will come from those who hold to the Mosaic account of Creation. He is therefore intent on outflanking them.

Thirdly, and most important, he makes it clear that he, with others, was scheming to propagate their "march of honest feelings" in a disguised way over some thirty years. What could these views be but the Theory of Evolution, which he had prepared the way for by his book and in which he had given much evidence in its support?

Just how close Lyell was to his two friends, Stokes and Broderip, is evident from the fact that his rooms at Grey's Inn were on the same staircase as Broderip (who possessed a large collection of shells) and that Stokes was very near [27p304].

It is indeed characteristic of Lyell that in his propagation of the false theories of Uniformitarianism and Evolution, he should camouflage them with the label "honest feelings ".

Darwin's arrival on the scene

From what I have recounted above, it will be obvious that Lyell's main objective was *not* just the formulation of a geological theory, but the overthrow, by devious means, of the Genesis account of the Deluge, with the further intention of preparing the way for the propagation of evolution. Far from Darwin being enabled to "work out" "his theory" of evolution in the light of the vast periods of time which Lyell's Uniformitarianism had provided, the truth was quite otherwise. Lyell and his collaborators had already set the stage and provided the "props".

Only a year or two after he had written this letter, he heard of a young naturalist touring the world who was sending home some remarkable reports and specimens, and he determined to make his acquaintance. Darwin returned from his voyage on the *Beagle* at just the right moment to be chosen for the next act in the play. Lyell's prediction of "thirty years' march" from his letter of 1832 was to prove remarkably accurate when that same young explorer was to eventually publish "his" theory of evolution in 1859.

DARWIN AND LYELL

Darwin landed in England in October 1836, and Lyell seems to have lost no time in making himself a close acquaintance of Darwin, for shortly after his arrival he wrote to Henslow, saying:
> "Mr. Lyell has entered in the *most* [emphasis his] good-natured manner, *and almost without being asked, into all my plans* [emphasis mine]".

Similarly, in April 1837 he said he was flattered by the interest of "Lyell, who has been to me, since my return, a most active friend" [2 p280].

Wilson also confirms just how keen Lyell was to meet Darwin and influence the course of his future activities as will be seen from the folowing quotations -
> "How I long for the return of Darwin" (Lyell in a letter to Sedgwick)[27p425].

> "You cannot conceive anything more thoroughly good natured than the heart and soul manner in which he put himself in my place and thought what it would be best to do" (Darwin in a letter to Henslow)[27p434]

Wilson furthermore notes that
> "...Lyell also offered to read some of Darwin's rough papers."[27p434]

and that
> "...Lyell and Darwin continued to see each other frequently." [27p457]

Thus Lyell appears to have made himself Darwin's close friend and confidant. I would suggest that this was done deliberately with a view to influencing the career of this rising young naturalist. Darwin would doubtless be greatly flattered to have so eminent a figure as Lyell as a personal friend. Lyell, on his part, would be willing to feed Darwin's vanity as illustrated by a letter he wrote to him in

1841, in which he said:

"It will not happen easily that twice in one's life... a congenial
soul so occupied with precisely the same pursuits and with an
independence enabling him to pursue them will fall so nearly in
my way, and to have them snatched from me with the prospect of
your residence somewhat far off is a privation I feel as a very great
one".

Lyell was keen to bring Darwin into his own circle of friends,
and in a letter dated 26 December 1836 (in which he has clearly
vetted a paper by the young Darwin), he invites him to dinner
saying: "...one or two are to be here, to whom I should like to
introduce you, besides a few whom you know already". Lyell was
also the prime mover in obtaining Darwin's admission to the
prestigious Athanaeum Club.

Just how much Darwin relied upon Lyell for guidance in much
of what he did is seen in some of his letters, such as:

"... It is certainly true that I owe nearly all the corrections [in
the second edition of 7 January 1860] to you, and several verbal
ones to you and others. I am heartily glad you approve of them"
[3p264].

Indeed one has a distinct impression that Darwin did very little,
either by action or writing, without first consulting his "tutor".

As we have seen, irrespective of Darwin's later claims, his first
notes ever to refer to the "species question" were not made until
July 1837, some ten months after his return to England. What could
have caused him to take such an interest in the subject?

Bearing in mind all that I have set out above, I would suggest
that *Darwin was induced to write about the theory of evolution by
Lyell*.

Lyell would have little difficulty in raising the topic in such an
oblique fashion, that Darwin would be quite convinced that the
whole of the idea would be 'his'. Lyell would praise him for 'the
boldness of his thinking' and his 'brilliantly clear reasoning' and
would infer that success in 'proving' 'his' theory would result in
worldwide fame - a possibility which Darwin would find irresistible.

One could indeed go further. It seems likely that Lyell's sole
purpose in rapidly befriending Darwin and promoting his interests
*was in order to use him as a means of propagating the theory of
evolution*, which Lyell was already determined to advance, despite
the seeming vacillation in some of his letters.

How aware was Darwin of Lyell's real intentions? During the
early days of their friendship this is difficult to determine, but it
would appear that initially he was more an innocent dupe. However,
in his later years he was seemingly not only aware of Lyell's aims,
but also, by his continued support, an active participant in the
scheme also.

In 1873 he wrote a letter to his son, George, (which was omitted from his published letters). In this he mentions that the impact of John Stuart Mill was much greater because he had not expressed his religious views and continued:

"Lyell is most firmly convinced that he has shaken the faith in the Deluge far more efficiently by never having said a word against the Bible, than if he had acted otherwise....I have lately read Morley's Life of Voltaire and he insists strongly that direct attacks on Christianity (even when written with the wonderful force and vigour of Voltaire) produce little permanent effect: real good seems only to follow the slow and silent attacks"[1 p320].

Seven years after he had written this letter, he expressed much the same view in a letter to Edward Aveling - Karl Marx's son-in-law saying:

"Moreover though I am a strong advocate for free thought on all subjects, yet it appears to me (whether rightly or wrongly) that direct arguments against Christianity and theism produce hardly any effects on the public, and freedom of thought is best provided by the ["gradual" added] illumintion of ["the" deleted, "men's" added] minds, which follow from the advance of science. It has, therefore, been always my object to avoid writing on religion and I have confined myself to science."[60p161]

Thus Darwin reveals that his purpose in "illuminating men's minds" with his theory was fundamentally an attack upon Christianity, and that furthermore he was well aware of the ultimate objective at which Lyell and others were aiming.

I will leave my reader to ponder upon the mentality and motivation of men who call secret attacks upon Christianity "real good"!

We have already shown how deeply laid were Lyell's plans, how long term his aims, and the cynical way he deceptively flattered the ecclesiastical opposition simply in order to destroy their present faith in the Mosaic account.

It may be objected that these are rather strong conclusions to reach, based upon a few selected passages in some of his letters. I, however, would maintain that, firstly, these passages allow us to see Lyell's real motive, i. e. he has momentarily allowed the mask to slip. Secondly, it must be remembered that these appear in an *edited* collection of his letters and one is left wondering whether others were omitted for one or more reasons of expediency. As we have seen, there are many important passages and complete letters of Darwin's which were omitted from his published volumes. Finally, no man as astute as Lyell is likely to write to his close confederates of his schemes in an ordinary personal letter, for there is always the possibility that they may be made public at some stage. Such matters would be a subject for discussion only — discussions which naturally would go unrecorded.

I would therefore contend that even though the evidence is not overwhelming (as one cannot expect in such matters), it is sufficient to show that it was the intention of several men, who succeeded to a number of influential positions, to overthrow by the most devious of arguments and stratagems the Biblical account of Creation and the Flood.

The Bible was simultaneously being attacked from another source- by the 'modernist' theologians of the 'higher critical' school using the same long-term tactics. How the general acceptance of the accuracy and reliability of the Bible was finally destroyed under these twin assaults is a matter of history well worthy of further investigation.

CHAPTER 23

EVOLUTIONISTS AND REVOLUTIONISTS - LINKED?

Were there forces even more sinister at work than just Lyell and a few others who were simply intent upon changing the views of the scientific world on the subjects of geology and biology? Those who seek conspiratorial forces at work would probably claim that there were. For myself, during the considerable amount of reading necessary for this book, I could not help noticing the number of connections existing between those who propagated evolution and those who were associated with revolutionary ideologies and activities. As there is a possibility of these connections being relevant, I list below some of those which I noted.

1. MALTHUS

Malthus' father,"a personal friend of the philosopher and sceptic David Hume, was an ardent disciple of Jean-Jaques Rousseau" [17] who made him his executor. This friendship was formed during the years 1766-1770 when Rousseau was in this country. He escaped from France to avoid arrest and on being expelled from Berne came to England at the invitation of Hume.

The writings of Rousseau, and Godwin after him,"took for granted the perfectability of mankind and foresaw a millenium in which rational men would live prosperously and harmoniously without laws and institutions". Malthus himself disagreed with this particular aspect of his father's views and as a consequence wrote his famous *Essay* in which he predicted quite the opposite - that mankind would live in poverty due to the continual struggle to obtain food.

It might be asked, what was the source of Malthus' view that the whole of nature is in continual warfare? Intererestingly enough it was Benjamin Franklin, for Malthus noted in the opening pages of his *Essay*:

"It is observed by Dr. Franklin that there is no bound to the prolific nature of plants or animals but what is made of their crowding and interfering with each others means of subsistence.[He then gives Franklin's hypothetical example of one plant overrunning an area where there were no other plants, and finishes-] This is incontrovertibly true..."[24p2]

Franklin was the American Ambassador to France and, during his stay there, he became friendly with Voltaire. Franklin's connection with subversive movements was recognized by George III, who considered him the "evil genius behind the American Revolution" [18p97].

Malthus had at least read Godwin, Condorcet and one or more books by Franklin and, with his father's close connections with Rousseau, Malthus was doubtless well versed in subversive revolutionary ideologies. His *Essay* was simply his own alternative view on one of the various ideas stated by such writers. The fact that he wrote only against one particular side issue suggests that he was not fundamentally averse to the basic aims, for had he been critical of his father's views in general, he would surely have made this more important topic the subject of his *Essay*.

To be fair however, Malthus warned the nation against "the rule of the mob", for he considered that a new government would be just as unable to feed the poor as the former. He then said:

"The effect of the revolution in France has been, to make every person depend more on himself and less upon others. The labouring classes are therefore become more industrious, more saving and more prudent in marriage than formerly."[24p31]

A doubtful proposition indeed! This wildly inaccurate statement, since shown to be quite false [36], is but one of many which have resulted in his theory being completely discredited.

As we have already mentioned, the adoption of his theories brought about the very conditions which were likely to generate unrest and revolution. Fortunately for England, the Great Awakening under the ministries of George Whitfield and the Wesley brothers in the eighteenth Century established a widespread sound Christian faith in the land to such an extent that it was able to resist the forces of revolution.

We might perhaps pause at this point and reflect upon the world outlook of Malthus and many of those who fiercely propagated the theory of evolution. For the vast majority of people, the world of nature is an endless display of beauty, design and amazingly complex interrelationships in which the shedding of blood amongst animals is acknowledged and accepted but not allowed to mar the beauty of the whole panorama. It is perhaps indicative of the mentality of Malthus, Franklin and most evolutionary writers that when they look at precisely this same nature they see not beauty but only "struggle", "survival of the fittest" and "nature red in tooth and claw".

One is reminded of the motto "Honi soit qui mal y pense" which may be rendered "Evil be to him who evil thinks".

2. ERASMUS DARWIN

Erasmus Darwin is also said to have met and corresponded with Rousseau [18]. The fact that Darwin's evolutionary ideas were deeply subversive of established religion was recognized by George Canning who was in Pitt's government. Darwin was also under suspicion for his connections with revolutionaries, both American and French, such as Benjamin Franklin [18 p97] and Thomas Beddoes.

As part of the suppression of anti-Church and State forces by Pitt, Canning published a parody of Darwin's *The Loves of the Plants* (which his biographer noted was "subtly subversive"), as a result of which Darwin's popularity declined rapidly [18 p265].

3. THE LUNAR SOCIETY

This was a society which met in Birmingham and consisted of some of the foremost engineers and inventors who brought in the industrial revolution. Amongst them were Josiah Wedgewood (the famous potter and great grandfather of Charles Darwin's wife), Erasmus Darwin, Matthew Boulton and James Watt. Although the main interest was the development of various mechanical devices, there does seem to have been a strong connection with the membership of another group known as the Revolutionary Society. A principal figure in both groups was James Kier, whilst Wedgewood and Darwin were known to be sympathetic to revolutionary ideas. A leader of the Lunar Society was a Dr. Small who had come from America with a letter of introduction from Bejamin Franklin. When there was a riot against the Revolutionary Society the houses of two of the members of the Lunar Society were ransacked, and—

"...for the Lunar Society the wreck was total, or very nearly so; for Darwin, the Birmingham riots were a clear smoke signal. Britains brief flirtation with the French Revolution was over.... From now onwards Darwin became much more cautious in publishing radical opinions"[18 p212].

Shortly after the French Revolution, the Revolutionary Society forwarded an *Address* to the French National Assembly which was taken to them by Earl Stanhope, an ardent supporter of the Revolution [42 p9].

Josiah Wedgwood once wrote to Darwin:

"I know you will rejoice with me in the glorious revolution which has taken place in France" [18 p201].

To be generous to them, one can only hope that they were both under the popular (but mistaken) impression that it was an uprising of an oppressed peasantry instead of the machinations of subversive groups, regarding which few are aware [36].

4. CHARLES LYELL

Lyell came from a Scottish estate bought by his father who was a self made man. The revolutionary activist Gabriele Rossetti had escaped from Naples after the revolt of 1820 and had been brought to this country in an English flagship. He had written his *Comento Analitico* which was a commentary on Dante's *Divine Comedy* but giving it a political interpretation. Mr. Lyell read this work and "immediately wrote a warm and friendly letter to him". Thus began a friendship which lasted to the end of Mr. Lyell's life [27p188].

Mr. Lyell asked his son Charles to write an article for the *Quarterly Review* on the work, which he agreed to. He used all the legal skill he could muster in order to "raise the author... above the contempt into which his publication has thrown him amongst a large part of the literary world in town". A copy of Charles' proposed article however was sent to Rossetti who was outraged at the tone of the review. Lyell also seems to have had second thoughts for he considered that such an article may well do harm to the *Quarterly Review* and therefore decided to withdraw it. This whole incident seems to have soured the relationship between Charles and his father for the latter complained to Rossetti of his son's "ill natured sneers and sarcasms" [27p188].

In the course of his geological travels, Lyell made several visits to Paris where he seems to have had immediate contacts with all the top French scientists of the day.

During these visits he met Cuvier and Laplace. He also went on tour with L.C. Prevoste, who had been a pupil of Lamarck.

Laplace is well known for his theory on how the earth evolved from matter expelled from the Sun. This theory, although now found on examination to be quite inadequate in explaining many facts, nevertheless still receives considerable publicity in evolutionary text books.

As I have mentioned, Lyell employed Gereard Deshayes, who possessed a large collection of sea shells, to make a catalogue of them for inclusion as evidence in Lyell's *Principles of Geology*. I have already referred to the problems involved in determining the various species of sea shells. With such widely different views, Deshayes would have little difficulty in so classifying his collection to provide the necessary evidence to support the Uniformitarian Theory, which both he and Lyell were keen to propagate. They were both ardent supporters of Lamarck's evolutionary theory, for Deshayes had edited Lamarck's works, whilst Lyell's admiration for Lamarck is clear from a letter he wrote to a friend, in which he said:

"That the earth is quite as old as he supposes has long been my creed and I will try before six months are over to convert the readers of the *Quarterly [Review]* to that heterodox opinion" [27p161].

5. CHARLES DARWIN

Darwin, living a sheltered life at Downe, deliberately avoided the subjects of politics and social reform. The few contacts he had with Dr. Edward Aveling, Marx's son-in-law, are however of interest.

Marx, a great admirer of the writings of Darwin, sent him a copy of the second edition of his *Das Kapital*. Darwin in his letter of thanks written in 1873 said:

"...Though our studies have been so different, I believe that we both Earnestly desire the extension of Knowledge, & [that added] this in the long run is sure to add to the happiness of Mankind."[63]

Darwin wrote another letter in 1880 which has been used for communist propaganda for it was said that he wrote it to Marx turning down an offer to dedicate an issue of *Das Kapital* to him. It has since been shown however that it was to Dr. Aveling and that Darwin was refusing to have a book dedicated to him which Aveling had written. It is the extract from this letter which refers to "direct arguments against Christianity and theism produce hardly any effects" which is quoted in the preceding chapter.

Darwin was very careful to avoid becoming involved with political or religious controversies; he was obviously aware that his theory of evolution had created more than enough notoriety for him to deal with. In particular he was keen to disassociate himself from Aveling and his colleagues, Charles Bradlaugh and Annie Besant, as their conduct and social policies were notorious in Victorian society.

Close friends of Aveling - who included Bernard Shaw - regarded him "as a scoundrel" and he was "utterly unscrupulous as far as money and women were concerned". His common-law wife was Marx's daughter and she commited suicide, "her spirit broken by long experience of Aveling's indignities and dishonesties". It was suspected that Aveling, who could forge his wifes handwriting, forged her name on the order for the poison which killed her.[64]

Darwin did however agree to see Aveling one year after he had written to him. Aveling came to Darwin's house accompanied by Dr. L. Buechner, the German materialist philosopher and President of the Freethinkers Congress when the latter visited this country in September 1881. R. Colp describes this saying:

"After luncheon Darwin conducted Buechner, Aveling and his son [Darwin's son Francis] to the privacy of his study. On Darwin's insistence they talked about religion. Buechner and

Aveling stated the Freethought position. Darwin then smiled and repeated to his guests the same thoughts that he had written to Aveling one year previously:"'Why should you be so aggressive? Is anything gained by trying to force these new ideas upon the mass of mankind? It is all very well for educated, cultured, thoughtful people; but are the masses yet ripe for it?"'. Darwin then stated that since the age of forty he had ceased to believe in Christianity, because it was 'not supported by evidence"' [64p390].

Aveling wrote a pamphlet in 1883 in which he described this visit. Francis' read this and rebuked Aveling for claiming that "[Darwin's] Agnosticism was only Atheism writ respectable" [1p319]. Reading Francis' criticism of Aveling [2p317] it could be inferred that Francis was *not* present at the meeting, in which case one would be relying on Aveling's word - a dubious proposition to say the least. However, as Francis makes no comment whether he was there or not we must presume that he was, otherwise he would surely have corrected Aveling on this important point. It must therefore be accepted that Darwin did make the statement reported by Aveling.

We have examined the reports of this meeting with some care as Darwin's statements conflict with accounts that Darwin returned to the Christian faith in his last years - a subject which we consider in some detail in Appendix I.

Turning to another topic, we have noted that Darwin met a number of prominent historians of the day. Often these meetings took place in the home of Lord Stanhope. Darwin recalls that he occasionally dined with the old Earl Stanhope and in his autobiography he said:

"He seemed to believe in everything which was to others utterly incredible. He said one day to me 'Why don't you give up your fiddle-faddle of geology and zoology, and turn to the occult sciences?'" [2p76]

The Earl Stanhope he is referring to was the son of the same Earl Stanhope we have mentioned above, who was a strong supporter of the French Revolution. That the Earl Stanhope whom Darwin met should be deeply involved in occult practices raises interesting possibilities.

6. A.R.WALLACE

On his return from Malaya, Wallace was quickly accepted as one of the scientific elite of his day and wrote extensively on a variety of subjects. He was an avowed socialist and wrote many papers seeking to rectify some of the appalling conditions of the poor which existed in Victorian times. He contended that one of the most important causes of poverty was the ownership of land and he was a keen supporter of the Land Nationalisation Society of which he became President.

He realised that it would take many years to have any effect in this area, so he also tried to change the publics attitude to an even more difficult matter - that of the use of interest on capital by large monopolies. In 1884 he wrote a paper in the The Christian Socialist entitled The Morality of Interest - The Tyranny of Capital. What is interesting is that he had clearly read and approved of the writings of Karl Marx for he said:

"By the methods here sketched out the labourer will receive, as Karl Marx and other social reformers maintain that he should do, the whole produce of his labour, and he will obtain this general result without any aid from Government except...removing the restrictions on freedom which now hamper him."[52p248]

How was this to be brought about? Strangely, not by drafting laws which would limit the amount of capital which could be accumulated, *but by abolishing laws against it!* He says "Without any laws against usury, usery will practically cease to exist "! Convoluted nonsense such as this, set out with such confident phrases as "it is evident that" and "as all experience shows" is the hallmark of theoretical economists who purvey impractical social reforms under the guise of "freeing the labouring classes". Just how deeply Wallace had drunk at the fountain of Marxist teaching is shown by his call to the government to "free" the labourer saying: "But first unloose your bonds and cease to hamper her with your legal meshes, and then see if she will not achieve a glorious success." This call is clearly reminiscent of Marx's well known slogan which apppeals to the working classes to rise in revolution "for you have nothing to lose but your chains".

Whether we may agree with Wallace's political solutions or not, his apparent concern for the degrading poverty of many of the working classes is at least to be commended for the conditions of the Victorian poor were absolutely appalling. But was he really sincere in what he preached and did he put his principles into practice? It does not take much reading of his autobiography to discover that in two areas at least his private actions were at variance with his public protestations against the existing order.

A. *Investments*

In the article referred to above he urges that a number of fundamental reforms should be put in hand:

"...all loans should be *personal*, and, therefore, *temporary*" [emphases his] and should be provided by the local community for its own development;

"To this end the first step would be to get rid of all Government funds, *guaranteed loans, rialway stock*, etc., which are the main agents and tools by which capital is accumulated and money is made to breed money."

In another article entitled *A Substitute for Militarism* he waxed very emotional in discussing the way in which the Government exploited the Colonies declaring:

"We profess religion. We claim to be more moral than other nations, and to conquer, and govern, and tax, and plunder weaker peoples for *their* [emphasis his] good! While robbing them we claim to be benefactors !"[52 p224]

He pays special attention to India and claims that it is:

"...a country which we rule and plunder for the benefit of our aristocracy and wealthy classes..."

Thus in his writings at least, he opposed large investments, exploitation of the colonies, guaranteed loans and railway stock as all these simply kept the idle supplied with wealth which they did not earn.

But what happened to the money which he made from the sale of his valuable specimens collected in Malaya ?

It was invested in Indian guaranteed railway stock which gave him a very comfortable income of £300 per year. Thus he diplays just how deep were his much vaunted 'socialist principles'. Were it not a serious matter, Wallaces 'principles' are at such variance with his actions that it is almost laughable. But this is not the end of his fortunes.

He admitted that with the sale of the remainder of his collection he would have had an annual income of £500 — certainly more than sufficient to make him a wealthy man for the rest of his life. But he was not satisfied with this. On the advice a friend he reinvested the money in a number of other foreign railway stocks in the hope that their value would rise. Gradually however their value fell and he was "almost ruined". Other investments in some new developments by friends also failed and by 1880 he said he was in financial difficulties.

Fortunately relief was to hand. He mentioned his situation to a woman friend who subsequently mentioned it to Darwin. Darwin raised the matter with Huxley who promptly wrote to Gladstone suggesting that Wallace be granted a state pension 'in recognition of his scientific work'! This was speedily arranged and in the following year (and presumably for the remaining 22 years of his long life) he received a pension of no less than £200.

One does not like to see anyone reduced in their financial circumstances, but it is interesting to reflect on several aspects of this whole affair. Although he had a more than adequate income from his original investments, Wallace nevertheless displayed considerable avarice in gambling it on the stock market in the hope of increasing it still further. This venture having failed, he was rescued from comparative poverty by adroit string-pulling so that he would be provided with a good income. There is the added irony that the source of the pension was from the very institution he despised - the Government - who provided this at the (working class!) taxpayers expense ! He obviously benefited greatly from being an accepted member of that nebulous body known as The Establishment. In later years he appears to have done very well financially from lecture tours in N. America and from his books but I somehow doubt if he displayed his public spirit by then refusing the state pension.

B. *The Poor*

As one may expect, in his socialist writings Wallace always expressed his strong feelings regarding the exploitation of the working classes. Yet again, just how deep was his concern?

He admits that:

"Up to middle age...I was so much disinclined to the society of uncongenial and commonplace people that my natural reserve and coldness of manner often amounted, I am afraid, to rudeness....Hence I was thought to be proud conceited and stuck up."[52p382]

Somewhat surprisingly, he claimed that it was his interest in spiritualism and phrenology (a study of the shape of the head to reveal personality) which made him realise that "there were no absolutely bad men and women" and that given equality of opportunity they could all become useful and happy members of society.

A careful reading of his very lengthy and detailed autobiography however provides not a single case of him mixing with or speaking to any of the poorer classes in order to help them in their distress. Indeed, on the contrary, he appears to have circulated only in the very highest of circles in the course of his lectures on evolution and meetings in private houses amongst the wealthy of the land when attending numerous seances.

Perhaps it is significant that the *only* reference he makes to the experiences of the very poor are two incidents related to him by a well known Christian. The Rev. Hugh Price Hughes was telling Wallace about his work in the London slums and gave them as examples of how sincere Christian love, when shown to people, can transform even the most degraded of personalities.

Anarchism

Wallace's claim to be a convinced socialist does leave it an open question of precisely how far he was prepared to take the restructuring of society. We can gather some idea of how extreme his views were from a discussions he had with a famous French Geographer Elisee Reclus who visited him about 1881. He writes:

"However, we did not talk of geography during the afternoon we spent together, but of Anarchism, of which he was one of the most convinced advocates, and I was very anxious to ascertain his exact views, which I found were really not very different from my own. We agreed that almost all social evils - all poverty, misery, and crime - were the creation of governments and of bad social systems; and that under a law of absolute justice, involving equality of opportunity and the best training for all, each local community would organize itself for mutual aid, and no great central governments would be needed, except as they grew up from the voluntary association of their parts for general and national purposes."

"On asking him if he thought force was needed to bring about such a great reform, and if he approved of the killing by bombs or otherwise of bad rulers, he replied, very quietly, that in extreme cases, like that of Russia, he thought there was no other way to force upon the rulers' notice the determination of the people to be free from their tyrants; but under representative governments it was not needed, and was not justifiable"[52 p208].

Thus we see that Wallace, if nothing else, was prepared to work for the abolition of the present form of government and its replacement by a vague system of 'voluntary self regulation on a local basis'. He claimed that "no great central governments would be needed, except as they grew up from the voluntary association of their parts for general and national purposes". He clearly fails to see that, human nature being what it is, any such central organisation would quickly resume the role of overriding authority which all governments exercise today.

In his closing comment he notes with approval that Reclus "was a true and noble lover of humanity - a firm believer in the goodness, the dignity, and the perfectability of mankind". This concept of mankind gradually achieving perfection by its own efforts takes us back to the writings of Rousseau. It is this fundamental error which results in the failure of all human schemes for man's betterment. It is of course the very opposite to the Christian gospel which confronts men with the fact that there is no possibility of this being achieved, and that they must first humbly submit to Him who first created man perfect. It is little wonder that such a proposal, so damaging to man's self pride, is is so frequently rejected.

Many will dismiss the connections I have set out above as having no significance. Others who are convinced that conspiratorial forces are at work will read more into them. I must emphasise that all these points appear in books and articles which are freely available in any well-stocked public or university library.

The possibility of there being a connection between revolutionary groups and the theory of evolution has also been suggested by Dr. Henry Morris. In his book *The Troubled Waters of Evolution* he says:

"Although faith in the Bible and creation was still very strong in both Europe and America, resulting from the spiritual revivals of the Reformation Period and of the great awakening, there had been strong undercurrents of unbelief for a long time. Subversive revolutionary movements were influencing multitudes. Deist philosophers, Unitarian theologians, Illuminist conspirators, Masonic syncretists, and others were all exerting strong influences away from Biblical Christianity and back to paganistic pantheism. The French Revolution had injected its poisons of atheism and immoralism into Europe's bloodstream, and the German rationalistic philosophers had laid the groundwork for the destruction of Biblical theology in the schools and churches. Socialism and communism were on the upswing throughout Europe; Marx and Nietzsche were propagating their deadly theories and were acquiring many disciples - perhaps also financial backers, as students of conspiracies have frequently suggested. All of these people and movements were evolutionists of one breed or another" [47p59].

CHAPTER 24

THE SPREAD OF EVOLUTION

When Darwin first proposed his theory of evolution, it met with a barrage of oppositon. However, supporters of the theory, whilst comparatively few at first, nevertheless achieved positions of considerable influence and we can trace some of the ways in which these were obtained.

Lyell

Lyell deliberately avoided certain official appointments, as invariably they were time consuming administrative positions and distracted him from his main aim. However, he was keen to secure the position of Professor of Geology (although only an *amateur* geologist!) and a letter he wrote in 1831 reveals how he was assisted to achieve his aim. The position was in the hands of three bishops and two orthodox Doctors - one of whom demurred at his appointment, but "as Conybeare sent him (volunteered) a declaration most warm and cordial in favour of me, as safe and orthodox, he must give in, or be in a minority of one" [7p316].

How amused Lyell must have been that the description of him as "safe and orthodox" was accepted by the very men whose theological beliefs he flatly rejected but took pains to conceal.

In this same letter he quotes a colleague saying:

"Scrope writes, 'If the news be true, and your opinions are to be taken at once into the bosom of the Church, instead of contending against that party for half a century, then, indeed, we shall make a step at once of fifty years in the science - in such a miracle will I believe when I see it performed'" [7p317].

Although Lyell was not as prominent as some supporters of evolution were in obtaining positions of authority, he was neverthe-less ambitious to be a powerful influence behind the scenes. This is suggested by his own statement that:

"It is the seeing the superiority of others that convinces one how much is to be, and must be done to get any fame; and it is this which spurs the emulation..."[7p39]

Darwin also noted that Lyell –

"...was very fond of society, especially of eminent men, and of persons of high rank; and this overestimation of a man's position in the world, seemed to me his chief foible."[29p101]

Murchison

Another account of what could be interpreted as the infiltration

of the Establishment by various appointments is recorded in *Giants of Geology*. In describing Murchison's life, the authors note he –
> "...became a catalyst for British science in general. With his help young men secured positions, shy scholars came into official notice and associations which had been in the doldrums enjoyed crowded meetings. Now and then a critic muttered "meddler" or "politician", but his help was given with such generous goodwill that objections were very few" [10p109].

Interestingly Darwin's view of Murchison was very similar to his view of Lyell for he noted:
> "The degree to which he valued rank was ludicrous, and he displayed this feeling and his vanity with the simplicity of a child."[26p102]

We have already referred to Huxley's determination to "make a name" for himself and Darwin's criticism of Spencer's egotism. When the ambition and vanity of such men is considered, it stands in sharp contrast with the injunction that Christians should "do nothing out of selfish ambition or vain conceit, but in humility consider others better than yourselves".[Phil 2v3]

Huxley's influence

One of the most influential of all the scientists of his day was Huxley. Bibby notes that he exerted much of it by private contacts, and in addition he wielded considerable influence in the many important offices which he held. Bibby says he was a Fellow of four Societies, an Honorary Member or Fellow of nine Societies, was at one time President of five Societies and received honours from more than fifty overseas' organisations [12p3]. In addition, he was made a Privy Councillor.

By far the most influential group, however, was a small, restricted, virtually clandestine club which Huxley helped to form.

THE "X" CLUB

It was this little known group, founded in 1864, which exerted a profound influence upon the rapidly growing scientific establishments. Bibby says:
> "But perhaps the most important thing that Huxley did in 1864 was to help organise the X Club, barely remembered today but for nearly thirty years a powerful sub rosa influence on English science. Its nine members (Busk, Frankland, Hirst, Hooker, Huxley, Lubbock, Herbert Spencer, Spottiswoode and Tyndall) mustered between them a Secretary, Foreign Secretary, Treasurer and three successive Presidents of the Royal Society, six Presidents of the British Association, and several officers of the Geological and Linnean and Ethnological Societies. No tenth member was ever elected".[12p58]

Just how powerful this group was can be imagined, and

furthermore, Bibby mentions that they always dined immediately before meetings of the Royal Society, the most prestigious scientific body in the country and they were notified of a meeting of the club by the delivery of an algebraic formula. It is quite certain that the business of the forthcoming meeting was discussed and a policy mutually agreed regarding suitable subjects, speakers, elections, etc. Irvine confirms this, for he says:

"They discussed, often in a systematic manner, the politics of the learned societies, projects of new museums and journals, *the periodical warfare with religion* and the classics, the place of science in contemporary education" [16p183].

With a concerted approach being made by such influential men, they virtually controlled not only the business of the Royal Society, but many other important bodies also. By this means they would be able to determine the course which science in this country would take for many generations to come.

The Club, which Bibby describes as "gay and conspiratorial", functioned for no less than thirty years, eventually closing due to the increasing age and declining numbers of its members.

The influence of the 'X Club'

The Victorian era was the period during which industry and science developed at a prodigious rate. The need for an educated Middle Class and highly trained scientists meant that schools and departments in Universities were rapidly formed to supply this demand. Although Lyell was not a member of the club, it is interesting to read that "He also became a leader in the movement to reform British Universities."[10p96]

A group such as the X Club would obviously make every effort to ensure that as many posts as possible were filled by men who shared their views. There is little doubt that, as they were all convinced evolutionists to a man, *every single one of their appointees would also carry the banner of evolution to whatever positions they secured.* Bibby in fact lists some nineteen appointments which Huxley made for very senior positions in the newly formed universities.

As a result, it is not difficult to see how all the primary scientific posts would quickly align themselves with the new theory and eventually present a united agreement that "the evolutionary theory is proven by science".

This is still the situation prevailing today. Should anyone query *any* of the "proofs" of evolution, he is immediately ridiculed and referred to numerous books and opinions of "experts" which contradict his view, and he is laughingly dismissed from serious consideration, as being "against scientific evidence".

As in all human affairs, fragile theories are clothed with credibility and the subsequent course of history is changed, often for the worse, not because of the rightness or otherwise of the ideas, but because a small group of energetic men were prepared to work behind the scenes in order to ensure that their precepts were imposed upon society, with or without its consent.

THE OPPOSITION

Were there any groups who were prepared to counteract the influence of such organizations as the X Club in spreading the theory of evolution? One might have expected the main opposition to spring from the Christian Church. The Established Anglican Church, however, was in a low spiritual state and not organised to withstand the new "science" which boldly proclaimed, on the very thinnest of evidence, that the "scientific facts" provided "convincing proof" of the theory. Few clerics were sufficiently competent in science to criticise the "facts" presented, and in any case, to do so was to risk being made a butt of ridicule by the newly emerging scientific intelligentsia. Similarly, any moral or theological criticism against the theory would be presented as a case of "Science versus the Bible", with the obvious inference that Science was proven and true, whilst the Bible was only a fable and therefore, by implication, *un*true.

Of other denominations, the Brethren Movement held to the accuracy of the Bible, but formulated the Gap Theory. This proposed that the word "became" in the Authorized translation of Genesis 1 verse 2 meant that God destroyed a Creation which preceded Adam, and the earth "*became* without form and void". Into this "gap" they were then prepared to place the millions of years which geologists were claiming as the age of the earth.

This concept gives the impression that they were trying to hold on to the belief that the Bible was inspired and accurate, yet, by means of a poorly supported reinterpretation, at the same time avoid the stigma of being labelled "unscientific" and thereby lose credibility in the eyes of the public.

There are a number of objections to the Gap Theory, both theological and scientific, but some still today accept it as valid.

To my knowledge, the only other group who resisted - or rather ignored - the theory were the Free Evangelicals. Basing their doctrine and authority entirely upon the reliability of Scripture, they continued quietly on their way, ignoring the controversies being fought elsewhere. They were not immune from attack from another quarter however.

The 'theological' attack on the Bible

In the Eighteenth Century, Astruc, a pupil of Voltaire, pointed out that two different names were used for God in the Hebrew of Genesis 1 and 2 and throughout all the first five books of the Bible. He therefore claimed that the "tradition" that Moses was the author of them all was wrong and suggested that they were a combination of the writings of two authors. From this beginning sprang the whole of the theory of the multiple authorship of these first five books, which was later to crystallise into the Graff-Wellhausen Theory. The way in which the unjustifiable assumptions of the theory were loudly proclaimed as the "assured results of scholarship" so exactly mirrored the way in which the evolutionists boasted that their "facts" were "proved by science" that the similarity was more than remarkable.

Many theologians succumbed to the new "evidence", but Spurgeon and others, perceiving the principles at stake, opposed the tide of opinion. When his own church became embroiled in what became known as the Downgrade [of the Bible] Controversy, only he, with a few others, maintained their position.

As with the Universities regarding evolution, many theological colleges opened their doors to liberal scholars of 'Higher Criticism'. Indeed, Irvine, although calling it a "theological disaster", almost gleefully notes:

"At least partly concealed in the formidable Trojan Horse of Hegelian metaphysics, David Straus had got inside the evangelical fortress and surprised the garrison, subjecting miracles to such devastating textual criticism as seemed to convert the New Testament itself into a weapon against official religion" [16 p69].

Once these centres of training for future ministers had been captured, the spiritual decline of the churches was rapid, aided as it was by the lethargy and ritualism to which all large, human organizations are prone.

It is significant that Religious Faiths are now taught as having "evolved" from the primitive superstitions of early man. Even Jewish monotheism is said to have evolved from polytheism which they originally absorbed from other pagan tribes.

The counter attack from Bible believing theologians was slow in coming, but very thorough, as exemplified by the superb writings of the "incomparable Robert 'Dick' Wilson". Many books have been written showing the complete falsity of the Graff-Wellhausen Theory [e.g. ref 21], but they do not appear in the standard reference works in theological colleges or Departments of Religion in our Teacher Training Colleges. In one such College, the gift of a number of evangelical expositions of the Bible was refused by the head of the department but was eventually accepted after pressure from the lecturers.

Once again, the parallel way in which students are similarly shielded from any valid criticisms of the theory of evolution is more than coincidental.

OVERSEAS

I have only briefly considered how it was that all the major institutions in this country rapidly accepted the new theory. Overseas, the pattern was similarly repeated.

Germany

In Germany, Haeckel aggressively propagated the theory, playing the part of "Darwin's bulldog" in that country, which Huxley had performed in England. The acceptance of the theory in Germany was particularly rapid, the theory being referred to in the fields of sociology and economics as a "fact" long before such a stage was reached in this country. Haeckel's methods, however, are particularly dubious. His faking of drawings in order to bolster the now discredited Theory of Recapitulation have been documented in other works.[45 & 53]

America

The main protagonist in America was Professor Asa Gray, with whom Darwin corresponded shortly before the publication of his *Origins*. Irvine notes that:

> "Gray was magnificently equipped to lead the evolutionary crusade in the United States....[a man with] unparalleled power among his younger colleagues. He was a vivid and indefatigable writer of textbooks *and therefore a major force in education....* once aroused, he could wield the fork and net of controversy almost as subtly and dangerously as Huxley himself. Evolution gave him the opportunity for the *Origin* divided Harvard as it divided the world....It was during these years immediately after the appearance of the *Origin* that Gray joined the inner circle of those who shared the master's confidence and gave valued advice"[16p87].

The 'Scopes' trial

In 1922 Senator W.J. Bryan was campaigning against children being taught that they were descended from apes, and eventually went to court to fight the famous 'Scope's trial. While the case was in progress, a tooth was found in Nebraska which Prof. Osborn of the American Natural History Museum claimed had characteristics that were a mixture of human, chimpanzee and Pithecanthropus. He declared:

> "...the Earth spoke to Bryan from his own State of Nebraska. The Hesperopithecus tooth is like the still, small voice. Its sound is by no means easy to hear... This little tooth speaks volumes of truth, in that it affords evidence of man's descent from the ape."

In England, Prof. Grafton Elliot Smith had a double page picture of a reconstruction of Nebraska 'Man' and his wife printed in the London Illustrated News - all based on this one tooth!

Later, other teeth were found and it was admitted that they were of an extinct form of pig!. Naturally, little publicity was given to *that* discovery.

What is of particular interest however is the timing of this 'discovery' of the tooth which took place whilst the court case was still going on. I have no proof in support of my views but I cannot help suspecting that it was deliberately timed and given a great deal of publicity at that precise moment in an effort to sway the final outcome of the case. As it was, Bryan won, but was awarded a derisorily small sum as damages.

The 'Scopes' trial

In 1922 Senator W.J. Bryan was campaigning against children being taught that they were descended from apes, and pursued this through the courts. While he was engaged in this, a tooth was found in Nebraska which Prof. Osborn of the American Natural History Museum claimed had characteristics that were a mixture of human, chimpanzee and Pithecanthropus. He declared:

"...the Earth spoke to Bryan from his own State of Nebraska. The Hesperopithecus tooth is like the still, small voice. Its sound is by no means easy to hear... This little tooth speaks volumes of truth, in that it affords evidence of man's descent from the ape."

In England, Prof. Grafton Elliot Smith had a double page picture of a reconstruction of Nebraska 'Man' and his wife printed in the London Illustrated News – all based on this one tooth!

It was this tooth which was paraded a few years later at the famous 'Scope's trial' — a trial which proved to be a turning point in the propagation of evolution in America.

Later, other teeth were found and it was admitted that they were of an extinct form of pig!. Naturally, little publicity was given to *that* discovery.

What is of particular interest however is the timing of this 'discovery' of the tooth which took place whilst Bryant was compaigning against evolution. I have no proof in support of my views but I cannot help suspecting that evidence for evolution was badly needed by its supporters. Subsequently the tooth was 'discovered' and given a great deal of publicity at that precise moment in an effort to sway public opinion against him.

It is quite obvious that today every institution or establishment of importance and influence is firmly held to the acceptance of evolution, and all its teachings and research are carried out with the theory as a basic philosophy. With the takeover of the national institutions virtually complete, all effort is now directed to ensuring that contradictory views never receive any publicity.

CHAPTER 25

SUBSEQUENT FORTUNES OF THE THEORY

One of the major problems facing the evolution theory is to propose a mechanism whereby favourable characteristics can be acquired by species. Lamarck stressed that useful characteristics acquired during the lifetime of the organism could be passed on, an idea which Darwin accepted. He proposed that all the organs in the body gave off "gemmules", which collected in the germ cells prior to fertilization, giving the offspring the characteristics of the parents. This was the theory of Pangenesis (i.e. all the body contributed to making up the germ cells), but such theories were all speculation, there being no experimental evidence to support them.

The first truly scientific attempt to investigate why succeeding generations vary from the parent was made by an Austrian monk.

GREGOR MENDEL (1822-84)

Mendel conducted numerous cross-breeding experiments on pea plants and discovered that various characteristics (height, colour, etc.) appeared in first and second generation plants in accordance with simple arithmetic proportions. These rules of inheritance frequently appear in books dealing with evolution, often taking up considerable space in explaining dominant and recessive genes, alleles, etc. What is not often appreciated by the reader, however, is that there are aspects of the discovery made by Mendel *which seriously undermine* the use of this theory by evolutionists for they do *not* provide a mechanism for producing species with completely new characteristics.

Variation

The important aspect of Mendel's evidence is that the two parents already possess a variety of characteristics which will be passed on to the offspring. All that happens is that these characteristics are simply shuffled around when producing offspring, *and no really new factors appear.* Thus by selective breeding, certain features can be produced, but such "pure" breeds are still fertile with the main stock, i.e. they are still the same species. Furthermore, there is no possibility of such a mechanism producing a completely new type of animal. The numerous diagrams and descriptions of Mendel's work in books on evolution give the theory an appearance of being founded on careful scientific experimentation. Yet how many readers realize that the facts presented are *against* the theory?

Mendel's Results

One interesting (but not relevant) sidelight on Mendel's experiments is that the ratios and numbers of the various plants he gave were in some instances "just too good to be true". Genetecists have repeated his experiments and got the same broad results but not with the arithmetic accuracy which Mendel quotes. R.A. Fisher made a statistical analysis of his results and found that they were much more accurate than could have been normally expected. One set of figures would only have occurred once in twenty trials, another set once in four hundred and forty four, whilst the overall accuracy of his results would occur once in two thousand and trials [23].

Fisher's explanation for these too accurate results is that either Mendel consciously or subconsciously classified colours, height, etc. in the ratios he was expecting or an assistant provided him with plant counts he knew Mendel was hoping to obtain.

Whatever the reason, it does not invalidate Mendel's basic thesis which is well established. It does, however, throw an interesting light on how easy it is for scientists to obtain the results they are hoping for.

Mutation theory

That Mendel's results are against evolution has been acknowledged by genetecists and therefore another mechanism has been proposed, which has so changed the basis of Darwin's theory that it is now called Neo-Darwinism. This relies upon the fact that a *very* small percentage of offspring is noticeably different from the parents - these are called mutants. Their appearance is due to the genes of the germ cells being significantly affected. Badly damaged cells are aborted. Slightly damaged cells may live but possess some abnormality or disfigurement. It is these mutants which evolutionists claim may occasionally, under changing environmental conditions, be better adapted than the parents and thus propagate successfully. With further mutations, it is presumed that a new species will appear.

This theory has much against it. For example, it requires 110,000 generations to establish a mutation which gives a small advantage of 1 in 1000, assuming that conditions do not change. This is admitted by Colin Patterson in his book *Evolution* published by the British (Natural History) Museum [22p73], a publication incidentally which makes a number of important admissions against evolution theory.

Another objection is that in a population of, say, 10,000, this "chance" mutation would have to occur *independently* about *fifty* times for it to avoid being swamped out of existence by the parent stock.

Further criticisms could be made, but to set them out at length would require a complex technical appraisal. Suffice it to say that despite this, Darwinists have had to resort to mutations, for Mendel's experiments are so hostile to the theory.

MENDEL — LOST

Mendel obtained a degree at Vienna University and taught Natural Physics at the school in the monastery, where he rose to become Abbot. He was a member of a recognized scientific group whose papers were circulated to all the major European scientific bodies. Mendel published his results in a paper in 1866, but these were "overlooked" by the scientists of the day. Fisher says:

"The peculiarities of Mendel's work, to which attention has been called in the previous section [his "too accurate" results], seem to contribute nothing towards explaining why his paper was so generally overlooked. The journal in which it was published was not a very obscure one and seems to have been widely distributed. In London, according to Bateson, it was received by the Royal Society and the Linnaean Society. The paper itself is not obscure or difficult to understand; on the contrary, the new ideas are explained most simply, and amply illustrated by the experimental results" [23].

For whatever reason, Mendel's work was unacknowledged for thirty five years.

MENDEL — FOUND

In 1901 it was "rediscovered" by a Dutch botanist, Hugo de Vries, who claimed he had read the paper in the course of his research on mutations. What is a little strange is that two other workers, Correns and Tschermak-Sysenneg, are said to have also "rediscovered" Mendel at the same time and quite independently!

It was de Vries who first seriously proposed that it was the chance incidence of mutations in a suitable environment which eventually produced a new stable species. If this were so, then Mendel's laws of variations was irrelevant to the appearance of new species, for it was effectively by-passed. The mutation theory claimed it was now able to provide a mechanism for the generation of new characteristics which would be "selected" by a "suitable" environment.

What may not be appreciated is that De Vries theory appears to have been discredited. Nordenskiold says:

"...a Swedish naturalist, Heribert-Nilsson, carried out the entire experiment over again and came to the conclusion that the new generations of *Oenothera lamarckiana* [the plant on which

De Vries had experimented] only show fresh combinations of characters that already existed in the main species. De Vries ...denied the validity of [Mendel's] law as regards mutations...but this has been found to be a mistake. Consequently his theory on the formation of that plant collapsed."[29 p588]

The barrier of silence

It is generally thought that Mendel's results were 'overlooked' by the scientists of the day, the blame usually being laid upon the 'obscure' publication in which they appeared. Some writers however have admitted that the ommission was deliberate. On this subject Nordenskiold notes that –

"We have only to remember that Mendel denies variability in those characters that he observed, whereas all the biologists were just at that time seeking after variations as material in proof of natural selection; and then come these assertions as to absolutely constant or constantly divisible characters from the pen of a monk in a monastery! It would certainly have been a miracle if they had found support from the generation that had been brought up on Haeckel's 'Natural History of Creation'".

It would seem that Mendel's work was deliberately ignored for the simple reason that it did not support evolution. As Wallace admitted:

"On the general relation of Mendelism to evolution I have come to a very definite conclusion; that is that it is really antagonistic to evolution".

Furthermore, the way in which his work was 'rediscovered' suggests that it was only publicly acknowledged *when an alternative, pro-evolutionary mechanism had been found, i.e. mutations*. Only then could his theory be given any recognition, and when it was, popular books presented it in such a way as to imply that it provided scientific *support* for evolution.

Why was Mendel's paper never discussed when first published in 1866? As Fisher has said, the article was received by the Royal Society and the Linnaean Society. Surely some English expert may have wished to raise the subject, either in a written article or formal debate. It is here that the importance of the 'X Club' becomes apparent. It is not difficult to imagine the members of the club, with all the influence they wielded between them, making sure that papers so damaging to the theory of evolution were quietly turned down or prevented from reaching a wide audience.

In a similar fashion, Fisher recounts how von Nageli, a German biologist with whom Mendel corresponded, not only ignored Mendel's results but actually warned students *against paying too much attention to them*! The Encyclopedia Britannica provides some additional interesting information, for it mentions that

Mendel carried out further experiments with the hawkweed plant. Unfortunately they did not corroborate the results of his first experiment, as this plant is effectively self-fertilized. What is disturbing is that he was encouraged to experiment on this plant *by von Nageli*! It would appear that as well as ignoring Mendel's work and warning others off it, Nageli seems to have deliberately induced Mendel to enter what he guessed was a blind alley in order to throw some doubt upon Mendel's original discoveries.

With the barriers of silence imposed upon Mendel's work, the subject would be allowed to rest until a 'satisfactory' mechanism could be proposed. Only when the mutation theory appeared on the scene was the 'skeleton' of Mendel's work allowed to emerge from the well stocked closet of the scientific establishment.

CHAPTER 26

LATER DEVELOPMENTS

When mutations were first considered, it was confidently expected that they would provide a perfectly adequate mechanism which would allow new species to appear. It was not long, however, before serious problems arose. As we have said, mutations may be considered to result from damaging the genes. The main objection was that as mutations were rather infrequent, fairly large changes were necessary in order to explain how new species appear in a very short time. However, large changes were so detrimental to the organism, that it was either aborted, lived only for a short period, or was deformed and lived with a much lower metabolic rate.

On the other hand, if only a small mutation occurred, there was simply not enough time available for it to become "fixed" in the population, whilst it would take numerous "chance" mutations to produce a new organ or species. Even with the millions of years which evolutionists insisted on claiming, the period within which many new species appeared was too short for the large number of generations necessary.

This fundamental dilemma is openly admitted by Patterson who says:

> "So selection theory is trapped in its own sophistication: it asserts that small differences in fitness are effective agents of evolutionary change, yet differences of that order are not detectable in practice" [22p69].

As this problem became more obvious, many scientists began to realize that there was *no* satisfactory means by which new species could arise. As a result, during the period 1910-1930 *the theory of evolution suffered a serious decline in academic circles.*

It was Fisher who then carried out a mathematical appraisal, and tried to show that *theoretically*, under ideal or unchanging conditions, it was just possible that mutations *might* produce a new species. However, as we have shown, in practice any small change would be very unlikely to gain a foothold in a reasonable number of the parent population.

In view of this, Professor Goldschmidt has proposed that new species are the result of large mutations which are not lethal, i.e. they appear on the scene as "hopeful monsters"! No adequate evidence for this theory is provided, and it has been criticized by Patterson [22p149] and others. The fact that reputable scientists have had to resort to such ridiculous speculations demonstrates how desperate they are to provide a solution to the problem. Clearly, they have had to scrape the bottom of the evolutionary barrel very clean!

RECENT DISCOVERIES

It is generally taught that it is the genes of the fertilized cell which completely determine what the final organism will be. It is assumed that if the genes and chromosomes could be artificially manipulated, then it should be possible to produce a new species. However, it is now known that the way in which the genes control the various characteristics is exceedingly complex. Indeed, just one particular feature may be determined by a number of genes located at various points and working in various ways. The usual picture often appearing in elementary textbooks of one gene controlling one characteristic is quite inaccurate.

There have been several experiments, however, into the subject known as cortical inheritance, which make a fundamental change to our understanding of how the type and variety of offspring are controlled. Basically these experiments, which we will be considering later, have shown that the enclosing membrane (cortex) of the germ cell has a profound effect upon the type (species?) of animal the fertilized cell will develop into, the genes simply controlling what variety (height, colour, etc.) its final form will take. This is an important discovery and we will first of all examine the subject of inheritance to show how it might solve the long standing problem of defining species and 'kinds'.

The Genesis 'kinds'

In Genesis Chapter 1 we read that each of the plants and animals are to reproduce "after their kind" i.e. one 'kind' will not become another 'kind' no matter how many generations may arise. It will be remembered that it was Linnaeus who attempted to determine what were the main groups of animals (or closely related species) that were limited by each of the various 'kinds'.

It became apparent that the determination of such groups was far from easy, one of the problems being the difficulty in defining a species. One simple definition would be that if two groups were crossed and produced fertile offspring then they could be classed as only varieties of the same species. However, there are a number of problems with this simple definition. For example, group A may be crossed with B and group B with C, but if group C cannot be crossed with A, then can they all be classed as one species ? This is a problem to both creationists and evolutionists from their respective viewpoints.

For the evolutionist, if you cannot define a species but claim (as some do) that all creatures form an unbroken, continuous series, then how can you say that "species vary" when definable species as such do not exist according to such a view? One of Darwin's most vigorous opponents in America was Louis Agassiz who was a pupil of Cuvier and became a professor at Harvard University where he founded the Museum of Comparative Anatomy. He pointed out as early as 1860 that-

"If species do not exist at all, as the supporters of the transmutation theory maintain, how can they vary? And if individuals alone exist, how can the differences which may be observed among them prove the variability of species?"[37p23].

This objection is still valid today as some evolutionist have admitted.

To the creationist, the discovery of a mechanism in any organism (mainly in the reproductive cells) which would effectively restrict variations or range of species within certain limits would strongly suggest that this was the controlling factor in deteriming the particular 'kind' the cell would develop into. The inability to define even a species thus poses a problem. This has been further compounded for it has recently been claimed that species which have been separated from the parent stock have developed differently. When they have returned to the same area they do not mate with the original members. This was in fact suggested by Darwin as the way in which new species developed although he had no evidence at the time. However, the fact that closely related species *do* not mate seemingly due to various changes of habitat, mating periods etc. does not mean that they *cannot* mate and give fertile offspring. This evidence needs to be carefully examined for it would be in the interest of evolutionists not to try too hard in showing that these species *can* mate even if it is by artificial insemination.

With all that has been set out above, it can be seen just how confused the situation is when the central point of the dispute i.e species, cannot even be defined with any certainty by either side.

Not infrequently, creationist claim that species may vary but cannot be changed into another. What creationists should be claiming is that the 'kinds' of Genesis will not change into another 'kind' no matter how long they exist. But if the definition of a species is difficult, then that for a 'kind' would be even more so. A solution to this problem has however been proposed.

Cortical inheritance

Dr. Arthur Jones has recently [38] suggested that it is the information which is contained within the cortex (i.e. the "cortical inheritance" referred to above) that determines which particular 'kind' of offspring the cell will develop into. He suggests that this subject should be studied in depth by geneticists using this idea as a working hypothesis. Here indeed is a potentially fruitful field of research which would greatly support the accuracy of the Genesis account of creation.

As with Mendel's experiments, cortical inheritance provides a large but nevertheless limited range of forms within which an organism can develop. Evolutionists on the other hand require a mechanism which could produce an infinite variety of forms which would permit the transformation of an amoeba into a man. Is this perhaps why the whole subject has been studiuosly ignored by our established institutions in the same way in which they treated Mendel's work?

This whole subject is very important and reference 39 lists all the research papers mentioned by Dr. Jones. These are given in the hope that workers in this field will be encouraged to investigate it further.

CHAPTER 27

STIFLING THE OPPOSITION

A number of people, among them several scientists who oppose the theory of evolution, whether it be on Biblical or scientific grounds or both, have formed groups, written books, given lectures and done all that they could to publicize the very considerable amount of scientific evidence against evolution.

How many do oppose evolution would be difficult to say, but it is certainly far larger than most people would imagine, and the number is rising rapidly. In order to acquaint the reader with some of the main organisations in both this country and overseas, Appendix VI lists some details of the more important groups. These are given in the hope that some may be prompted to join one or more of them.

In America, the efforts of the creationist groups and their many supporters are at last beginning to have an effect. In some States, a law has been passed that where evolution is taught, then 'scientific creationism' [the scientific evidence for creation without reference to the Bible] should be given an equal amount of time. This is being hotly contested in the courts by Civil Liberties and teachers organisations who claim it is importing religion into the classroom 'by the back door'.

In this country the creationist's case has rarely received any recognition. Quite recently, however, various articles have appeared seemingly in response to the growing numbers of those who support creation and the public presentation of the evidence. What is noticeable in such articles (or programmes) is that, although they may admit that there are a few "difficulties" with the theory, they will then proceed at some length to "explain" how certain objections can be overcome. What is certain is that at no time will a creationist be given a fair chance to present serious *scientific* objections to the theory. Any contributions made by a creationist will be carefully edited so that the only comments published will be those dealing with the general aspects of evolution such as the theological problems of origins. Serious scientific evidence receives no publicity whatsoever for to do so may provoke awkward questions on whether the theory is factually true when for years it has been paraded before the general public as "no longer a theory but proven fact".

Numerous lectures have been given on the scientific evidence which contradicts evolution and supports Genesis. Often the question is asked why such facts are never shown on any television programme. The answer is simply that no matter how frequently or by what means the B.B.C. are approached, the end result is always a refusal to grant creationists a fair chance to present their case. The response is the same from any of the radio authorities, larger publishers, national newspapers, etc. etc.

It may appear that such a jaundiced comment is only to be expected from a minority group who fail to get a hearing for their peculiar views. However, all who lecture and write against evolution can testify to the very considerable interest which the subject generates *when people are presented with facts which they have never heard before*, whether Christians or not.

Such is the freedom — some would call it licence — in this country to discuss openly such subjects as adultery, abortion, euthanasia and other contentious matters, but the one subject which is sacrosanct is the theory of evolution. Such a statement may appear exaggerated, but there are numerous incidents which justify such a claim. Indeed, this is an issue of importance and worth further investigation. I therefore set out just a few of the many instances which have come to my notice, or have happened to me personally. In what follows, I have refrained from deliberately searching out sensational stories, but I am sure there are many such worthy of further investigation.

1. THE B.B.C.

A) DR. GISH

As a Director of the Institute of Creation Research, California, he has travelled extensively, speaking on the creation versus evolution debate. Whilst touring in this country, he said:

"When a creationist is available they'll put him on TV. I've been interviewed many times on television in the States. I've been interviewed on nationwide tv in New Zealand and Australia. But your BBC...I think they are the worst when it comes to excluding an alternative point of view"[41].

B) "LIFE ON EARTH"

This was an excellent TV. series about the world of Nature. Nevertheless, with the narrator, David Attenborough, frequently referring to "millions of years", and the "survival" and "emergence" of new species, it was really a thinly disguised vehicle for evolutionary propaganda. From speaking to various groups, I am certain that the BBC and David Attenborough received numerous letters of complaint about the distortions and speculations in the programme. Yet the existence of a large body of people who object to the main thrust of this programme, and others like it, is never acknowledged.

C) "IN THE BEGINNING"

This was a programme transmitted on BBC 4 on 14th July 1981 in which both evolutionist and creationist took part. The programme was keenly anticipated by many creationists as providing an opportunity for their case to reach the general public. After the programme, many felt a little disappointed in the way in which the creationist taking part presented their case. Each of them however was interviewed for for about three quarters of an hour, the majority of which obviously had to be reduced for the final programme.

It seems certain that any really damaging statements against evolution were omitted whilst those which were more innocuous or weakened the creationist position by an admission of any kind were duly transmitted. In one case, the opening question of the interviewer was "What particular statements do you want to make?" Needless to say, the points then given were not included in the final broadcast. In another case, as soon as the inadequacies of radiometric dating was referred to, the subject was dismissed by the interviewer, as being "too technical for him to understand. "

The broadcast, far from being a presentation of their case by creationist's, was simply what the public were allowed to hear when it had been passed through the evolutionary filters of those responsible for editing the programme. Having made the programme, the BBC will now claim that the creationists have been given 'a fair chance'. Had the broadcast been live, the creationist would probably have been far more satisfied with what was said on their behalf.

D) "GENESIS FIGHTS BACK"

The article about this programme in the Radio Times, which covered two pages, was a sympathetic description of the activities of Dr. Monty White, a well known creationist speaker. One gained the impression that the whole of the programme would be devoted to his work and it raised the hope that perhaps at long last the BBC were prepared to give a creationist a fair hearing. Such hopes were short lived, for the the TV programme, broadcast on 22nd November 1981, was just as much a vehicle for evolutionary propaganda as previous transmissions.

In a programme lasting 35 minutes, Dr. White appeared for only 3 minutes, there was an extract of a lecture by an American creationist, an American father expressed his concern about the way in which his son was indoctrinated with evolution at school, whilst Dr. Colin Patterson of the British Natural History Museum admitted that it was only when he (temporarily !) lost his faith in

Darwinism that he realised that the theory was held as a *faith* by biologists. The whole of the remainder of the programme was both an attack upon the creationist view and yet another vehicle for the latest variation of evolution theory known as 'punctuated equilibrium' - of which more anon.

The creationists were accused by Prof. S.J.Gould of 'wilfull distortion' and 'lying' about the new 'evidence' he has provided, Canon J.A. Baker of Westminster made some very inadequate comments on some Bible passages which he claimed creationists ignored or misinterpreted, whilst Prof. John Maynard Smith laughingly dismissed creationism as not being a science.

What was particularly noticeable was that all these comments occupied the whole of the second half of the programme and the views purported to be held by creationists was presented *by evolutionists* in such a way as to be a travesty of the truth. No creationist answer to these charges was allowed. Had such an opportunity been given, these quite false accusations could have been easily demolished.

Regarding the publicity for this programme, one is left wondering why the *Radio Times* should have given so much prominence to a creationist. The publicity would doubtless have generated a large audience. The programme itself however was heavily slanted to present the evolutionists case.

2. LOCAL RADIO

In 1978, Professor E. Andrews and I were interviewed by Rev. M. Hall for his weekly programme, "Christian Forum" on Radio Trent. When this was duly broadcast, it generated considerable interest. A few months later it was also used by the London Broadcasting Company, and again it provoked interest. In January 1980 the LBC re-transmitted the programme and used it to measure their "audience rating".

The response this time was "the largest they had ever received" [consisting of *twelve* letters incidentally!] and I was invited by the interested producer to have an interview one Sunday afternoon for transmission later that evening. After we had met, he mentioned: "some people actually believe the world was created in six days". When I replied that I also believed this and gave lectures on the scientific evidence in its support, his lower jaw almost dented the floor! The interview, in which I challenged evolutionists to a debate, was eventually completed, and, a little to my surprise, actually transmitted.

I am quite certain that this programme generated even more interest than the earlier one. Yet despite writing, telephoning and leaving messages, I have not received any response whatsoever, neither have they forwarded any letters requesting further scientific information, or details of my book on Ape-Men.

As a sequel to this, I subsequently met the Chairman of the Religious Advisory Board of LBC. I had given a lecture during which I had referred to this incident. He said he would try to arrange a debate on the subject and I suggested the name of a prominent Christian theistic evolutionist. The offer however was not accepted.

This sudden silence of the LBC interviewer following the broadcast is quite typical of an initial great interest shown by producers and editors, only to be followed by a sharp "cooling off". It seems to me that either they become aware of how 'hot' the subject is and fear ridicule from their professional colleagues, or alternatively (or additionally) they receive a direct order from senior executives that no further time or space is to be given to the topic.

3. THE PRESS

The same response is received when approaches are made to the National Press, periodicals, liberal Church newspapers, etc. Invariably, no matter how many letters may be received by a publication in response to a statement perhaps by the "Science Correspondent", whilst a few may be printed which criticize evolution in a general way, none will appear which give factual refutation. The usual excuse is "lack of space".

4. UNIVERSITIES

A) *Mr. Andrew Loose M. Sc.*

Mr. A. Loose took a B. Sc in astronomy. The subject he chose for his dissertation was "The Age of the Universe". In researching for this, he examined all the various methods which had been used to give a date to the earth, the solar system and the universe. In every single case he found that considerable speculation was involved, and effectively the results were made to comply with evolutionary presuppositions regarding time scales.

When in due course he presented his paper, his was the only one which the Professor of the Department personally attended. Shortly afterwards, an instruction was issued that in future students would not be allowed to choose their own subjects, but must first confer with and receive the approval of the lecturers.

He subsequently went to another university to take his MSc. Further discussions on the subject of the age of the universe with a number of eminent astronomers — including the Astronomer Royal of the day — failed to show that his conclusions were in error.

B) *Dr. Arthur Jones*

Whilst he was an undergraduate, Dr. Jones was awarded a prize for an inter-faculty essay on Communism, in which he was openly creationist. When his Zoology Professor awarded him the prize, he told Dr. Jones that "No one who does not accept evolution will ever do research in my department". Undeterred, he nevertheless applied and, somewhat to his surprise, was accepted. Later, when he was being interviewed for his Ph.D., the assistant interviewer bluntly asked: "If we award you your Doctorate, how do we know you won't then use it to spread your views on creation?" The senior interviewer, however, interrupted and said the question was unfair and should not be answered.

During the research for his thesis, he read a paper by a biology professor who had investigated an area of biology and obtained results which seriously conflicted with a fundamental aspect of evolution. However, after six years, no further papers were published. Dr. Jones subsequently met him at a conference and asked him why he had published nothing further on the subject. He replied in confidence that he had been subjected to severe opposition because of his work in this field and so had turned to less contentious areas of research. An American Professor studying the same subject was called "Lamarckist" by other biologists — as we have said, in modern biology this is the ultimate term of abuse!

Difficulties facing students

Perhaps at this point we should consider the difficult position of children and students who are creationists, when they are taught evolution at schools or colleges.

Instruction in schools about evolution is invariably given as if it were a proven fact, but when teachers are challenged on this, the reply is that it is only taught as 'a theory'. But this aspect is so little referred to, if at all, that a class of young children will certainly absorb it as proven fact. How often children tell their parents that because "teacher said so" then it must be an incontrovertible fact! The absorbtion of evolution would of course be greatly strengthened *if not a single fact against evolution is presented.* The reason for this may be that the teacher may be completely unaware of any such evidence existing. Alternatively they may realise that to teach anything different to the standard set of notes on evolution would be to enter fields of contention which are not conducive to a quiet life! Those who take this 'easy way out' may like to reflect what a disservice they are doing to the generation they seek to serve so conscientiously.

At universities, the situation is even more difficult. A young student may disagree with the evidence for the theory being given by a lecturer, but may be unwilling to criticize him before the whole class and thereby incur his displeasure.

Similarly, when answering questions in examinations (not necessarily on evolution), to refer to evidence which contradicts the theory could make just that difference between pass and failure, should his paper be marked by an ardent evolutionist. I am sure that such could happen and at times does. Lecturers cannot claim to be *totally* free of personal views when dealing with such a fundamental subject as evolution. Indeed, it is not unknown for some people to become highly emotional when doubt is expressed about the theory.

Summary

The few examples which I have given above do indicate the variety of ways in which the dogma of evolution is carefully protected from public exposure of its scientific weaknesses. Enquiries amongst ones scientific friends would, I am sure, reveal many more instances.

In this chapter we have only briefly considered how some of our major institutions effectively block any serious criticisms of evolution from reaching the the general public. There is however one establishment organisation which so far exceeds these in its popular promotion of the theory that it requires a complete chapter to do justice to the investigation of its activities.

CHAPTER 28

THE BRITISH (NATURAL HISTORY) MUSEUM

This establishment, even more than the BBC, acts as the main mouthpiece for evolutionary propoganda. Founded in 1881, there is little doubt that it would have been strongly sympathetic to the theory of evolution for Huxley was one of the main proposers that such a museum should be built. So keen was he to see this take place that he canvassed many men of influence including the Prince Consort. Huxley became a Trustee in 1884 and with such a man closely influencing its course the museum would have a strong evolutionary bias both in those who were originally appointed to direct its operations and in the formulation and interpretation of its chartered aims and statutes. Indeed it could be inferred from the inclusion of the word "history" in its title that its main purpose was to tell the 'story' of how nature came to be as we know it today. i.e. how it evolved !

In recent years the museum has come under an increasing amount of criticism from many quarters. It was the museum's 'experts' of the day who fully approved the fossils found at Piltdown in 1912-16. It is now well known that in 1953 these were exposed by the museum staff as fraudulent. In the first edition of my first book I referred to the way in which the vital evidence of Mr. Essex regarding the identity of the Pildown hoaxer was given to the Museum's investigating team when the fraud was revealed in 1953. This information however was never published by the Museum. I also related the fact that a Mr. Kennard, a respected member of the Geological Survey at South Kensington had intimated to Mr. Hinton of the Natural History Museum that he knew the identity of the Pildown forger. Yet again, this information never appears to have been followed up by the investigators.

In the second edition of the book I related the accusation by Dr. Halstead, who at one time was on the staff of the Museum, that certain personnel in the Museum itself (paricularly Hinton) were involved in the fabrication of the Piltdown forgery when the fossils were found in 1912. Furthermore the way in which the *original* artificially stained fossils were guarded from close inspection even by top experts indicates that their fraudulent nature was known at a very high level! If nothing else, all this certainly suggests that there is something strange about the activity (or lack of it) of the museum's administrators.

In 1951 the Evolution Protest Movement wrote to the Trustees of the Museum expressing their concern regarding Professor Le Gros Clark's book *The History of the Primates* which had been published under the auspices of the Museum. The main point of their memorandum was that the Professor had excluded four fossils which showed that true men existed long before the apes which he considered were ancestors of mankind. A reply was received from Sir Gavin de Beer, a director of the Museum, in which he briefly dismissed the four fossils with a short paragraph on each. A detailed answer to this from the Movement received no further reply. An account of the correspondence is given in pamphlet No.19 of the Creation Science Movement.

THE NEW EXHIBITIONS

In the last few years there has been some major criticisms of the way in which the Museum is going, this time however not from creationists but from highly qualified evolutionists, both within and without the Museum itself. The whole subject came to a head at a syposium on Vertebrate Palaeontology and Comparative Anatomy held at Reading University in 1978.

At the meeting, Dr.R.S. Miles, Head of the Public Services Department of the Museum outlined the reasoning behind the new series of exhibition sections, and a summary of his talk appeared in *Nature* on 26th October 1978 (v275 p682). From this we learn that a survey of visitors to the Museum showed that "the 'most common' age groups are children below 11 years of age". Strangely, although he gives the percentages for other age groups, in the article at least the figure for this particular group is not quoted. However Dr. Miles, on the basis of this survey, made the surprising statement:
"There is therefore some justification for regarding the
Natural History Museum as a 'kids' museum, for children under
11 years".
It is on this basic approach that all the new exhibitions have been modelled.

A paper was submitted to the Trustees in 1972 which recommended that a new department should be set up in order to change the image of the Museum by making it more attractive and instructive to the many young visitors who flock to the Museum each year. In reading Miles' extracts from this report and others subsequent consultative documents, it is not difficult to see how most of the proposed exhibits would be used to either teach evolution direct or by implication. I quote -
"[The existing exibition is] neglectful of natural processes....
the new exhibition should deal with all forms of life, with the
origin of the earth and its life....It would have a uniquely
comprehensive opportunity to display the full range of natural
history. The paper suggests that the content of the new exhibition

should be grouped under four headings: man, ecology, life process and behaviour, and evolution and diversity."

Of the other documents, that for 'Evolution and Diversity' stated -

"...the aim of the new exhibits on these topics should demonstrate the varied form and structure of species now existing, the historical relationships between them, and their status in the latest stage of continuous process of change on the surface of the Earth".

In due course the recommendations were accepted, finance granted and the Public Services Department specifically created to organise these new displays.

All the new exhibitions have followed precisely along the lines recommended in the original document. This has resulted in a distinct change in the main role of the Museum. From the original intention of it to act a a repository of knowledge concerning the world of nature, it is now becoming yet one more vehicle for indoctrinating the general public to accept the theory of evolution. Such a change however raised a storm of protest from a number of academics, and a reply to this approach appeared on the page of *Nature* following that which carried Dr. Miles' article.

The criticism

This article was written by Dr. B. Halstead of the Department of Geology at Reading University. Although an evolutionist, Dr. Halstead has written several interesting articles over a period of time in which he has revealed a number of strange incidents in the present day academic world which I am sure many evolutionists would have preferred not to have been publicised. In this particular article he implies that the newly created Department of Public Services of the Museum walked rough-shod over the established and reputable members of the various departments on the Museum staff. For instance he mentions that:

"One of the world's leading icthyologists on the museum staff was provoked to protest publicly at the recent removal of the fish exhibits to make way for the Human Biology Exhibition."

and that:

"The museum has announced its forthcoming publication on the new dinosaur exhibit and yet its own expert is not involved in its production. This in spite of the fact that as well as being acknowledged as a world expert, he is one of the most experienced people around in communicating his subject to young people and the general public"

Halstead continues by saying:

"Once it is conceded that it is an appropriate role for a national museum to be concerned with aspects of social engineering by promoting concepts that happen to be current in the present climate of opinion, then a most dangerous precedent is set which has sinister implications. Suddenly it becomes possible to visualise museums contributing to the indoctrination of the more inarticulate sections of the community. Just so long as natural history museums are primarily concerned with displaying the materials in their charge, there is always the possibility that the facts will shine throught the prevailing dogmas".

The new exhibitions are certainly strikingly different from the usual displays at museums for they have a considerable amount of "audience participation". Displays are arranged with coloured photographs or drawings and the visitor is invited to answer various questions by pressing a selection of buttons which illuminate the correct answer or give the reason why the answer chosen was wrong. Such activities are very attractive to the many young children who visit the museum often in large organized groups from the schools and they have clearly been deliberately designed for this particular age range.

Certainly the organisers of the various displays are keen to ensure that the visitors depart from the museum with considerably more knowledge about living organisms than when they entered. But what sort of information is it that they will mainly learn? A brief summary of the way in which evolution is subtly propagated in these exhibitions is given in Appendix II following a careful inspection of them at the time of writing (September 1981).

The main approach

With the very large sums of money made available to the museum to mount these displays, what an opportunity there was for them to show some of the amazing ways in which various creatures are designed and live their life cycles. For example there are the variety of different ways in which insects fly or hatch their young etc., etc. There are simply countless fascinating facts which could have been illustrated with animated diagrams and coloured displays. Yet what is it that the display organisers have chosen to emphasise? It is that of the *relationships* between the various species, which is another way of implying that they have evolved from a common ancestor without actually calling it evolution. So much space is given to this subject that there is much less space available to show some of these remarkable facts. Where some of the more simple interesting facts of nature are described, it is invariably claimed to be an 'adaptation' of a species to a particular environment.

There did not appear to be in any of the new exhibits a single example of the more astonishing features that we find in nature such as the design of various creatures or the relationships that exist between them. In many cases it is almost impossible to imagine how they could have evolved to their present state. An example would be the Bombadier Beetle which secretes a most complex explosive mixture which it uses to ward off predators. Numerous other examples could be quoted. Are these perhaps deliberately omitted because they might well make visitors wonder how such arrangements could possibly have been brought into existence by evolutionary processes ?

At the end of his article attacking the present trends in the museum, Dr. Halstead makes the surprisingly outspoken suggestion

"We can only hope that sufficient pressure can be brought to bear to curb the activities of the Public Services Department and to ensure the survival of the museum's reputation for scholarship in its public galleries".

To make a public call to stop a department of a prestigous organisation from continuing on its present course must be very unusual in academic circles. This however is not the end of the story.

Political motivation

In a lengthy letter which appeared two years later in the issue of *Nature* of the 20th November 1980 (vol 288), Dr. Halstead was even more outspoken in his criticism of the Department of Public Services in the Natural History Museum.

In the course of his letter he attacked two main aspects. The first of these is the frequent use of 'cladograms'. These are a way of showing diagrammatically the relationship between various species. I show the Museum's diagram for the evolution of man in Fig. 13

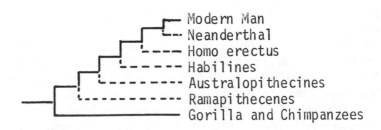

Fig.13. The Natural History Museum's Cladogram of 'Man's Ancestry'

The most noticeable aspect of such a diagram is that *not one of the fossils is accepted as being directly ancestral to modern man*!

As Halstead points out, the Museum openly acknowledge that they "assume that none of the species we are considering are ancestral to any of the others". Such diagrams allow the possibility that evolution proceeded by a series of 'jumps', and we only have the fossil evidence of the succesful species, the fossils of the original ancestors being 'lost' as it is assumed that they rapidly developed over a very short period of time.

This leads to Halstead's second accusation regarding why such diagrams were appearing in the Museum's exhibitions. I must admit I was very surprised to see in such an influential journal of the academic 'establishment' as *Nature* the blunt statement that the presence of these diagrams was due to revolutionary Marxist influence. I quote Halstead at length:

"...why should the notion of gradualism arouse passions of such intensity? The answer to this is to be found in the political arena. There are basically two contrasting views with regard to human society and the process of change through time: one is the gradualist, reformist and the other is the revolutionary approach. The key tenet of dialectical materialism, the world outlook of the Marxist-Leninist party according to J.V. Stalin, is in the recognition of "a development in which the qualitative changes occur not gradually but rapidly and abruptly, taking the form of a leap from one state to another" (Engels). This is the recipe for revolution. If this is the observed rule in the history of life, when translated into human history and political action it would serve as the scientific justification for accentuating the inherent contradictions in society, so that the situation can be hurried towards its appropriate "nodal point" and a qualitative leap supervenes."

"With regard to evolution and the fossil record, neither Engels nor Lenin, both of whom discussed the subject at length – to their great credit – insisted upon a pattern of such qualitative leaps, they were merely content to see in evolution and the fossil record evidence of change, albeit gradual."

"This has always been a matter of some disquiet for Marxist theorists. If it could be established that the pattern of evolution was a saltatory one after all, then at long last the Marxists would indeed be able to claim that the theoretical basis of their approach was supported by scientific evidence. Just as there are "scientific" creationists seeking to falsify the concept of gradual change through time in favour of catastrophism, so too there are the Marxists who for different motives are equally concerned to discredit gradualism."

"What is going on at the Natural History Museum needs to be seen in this overall context. If the cladistic approach becomes established as the received wisdom, then a fundamentally Marxist view of the history of life will have been incorporated into a key element of the educational system of this country. Marxism will be able to call upon the scientific laws of history in its support, with a confidence that it has [not?] previously enjoyed."

"This is the course of action to which the authorities of the Natural History Museum seem to have committed themselves either unwittingly or wittingly."

Here is certainly one of the most outspoken accusations made against the senior administrators of the Museum. As can be imagined, this letter stimulated a great deal of controversy, generating some thirty letters and three editorials in *Nature*. The staff and administrators of the Museum remained silent for some time until virtually stung into making a reply by one of the editorials in *Nature*.

With regards to Halsteads accusation of Marxist influence at the Museum, it might be pertinent at this point to refer to an interesting letter which appeared in *Nature* following the storm which Halstead raised. This was from H. Crabtree and noted:

"...the only Englishman at the graveside to hear this [Engel's oration at Marx's funeral] was Marx's erstwhile young friend Ray Lankester FRS. Lankester was later knighted and served as the Director of the Natural History Departments of the British Museum (1898-1907), and is still remembered for his reclassification of the Talpidae collection. The plot thickens!"[68]

From the accusations against the Museum I have made above and the strong reaction which the Museum's present activities have provoked from its own fellow evolutionists, it would appear that an investigation of precisely what is going on in that establishment is long overdue.

In his second letter, Halstead refers to his call two years previously for pressure to be applied to the Public services department but he makes no further calls. Does he perhaps sense that they are too late and of little avail? If this is so, then I would agree with him. I do not think such protests will have any effect whatsoever. I must confess that I am slightly amused at those, whether academics or laymen, theistic evolutionists or creationist friends, who express surprise when they fail to change the attitude of any of our institutions, particularly the Natural History Museum and the BBC, by writing and protesting to the various departments. I fear that they are labouring under the mistaken impression that they are dealing with men who are open to persuasion by the submission of reasonable views. I am certain that it is the intention of those in positions of authority *not* to be deflected from their aim. Indeed, I must admit that it is I who am surprised when there is some small acknowledgement that the creationist view even exists - but that is infrequent. I am convinced that all the major positions of power and editorial control are ultimately in the hands of those who for one reason or another will not allow any damaging evidence against evolution to reach the general public.

It might be worth pointing out at this stage the impregnable position an organisation such as the BBC is in. Even the Government of the day cannot affect the line they take for any interference is immediately labelled as 'political meddling with the freedom of speech'. Any pressure from creationists can be simply dismissed as 'fundamentalist backlash'.

Having made such a comment, nevertheless one can only hope that Dr. Halstead and his colleagues do manage to affect some change. I am however pessimistic. The surveys have been conducted, the recommendations proposed, the finance made available, the staff engaged and the whole operation put under way before the true nature of the intention has been realised and when protest is far too late. In other words, the cards have been carefully stacked against all opposition in such a way that they now form an impenetrable brick wall !

Summary

Doubtless the charge will be made that creationists are an obscure minority group who are clamouring to get their strange views propagated in the mass media. However, frequently it is the ordinary member of the public who asks why this information does not appear on television, etc. I find that lecturing on evolution produces considerable discussion and, were such programmes made, it would generate widespread interest.

Why is there such an embargo on the subject of the evidence against evolution? It seems obvious to me that the *whole* of the mass media in this country – and most others – is ultimately controlled by materialists/evolutionists, who carefully restrict criticism against their basic philosophy of life.

In England it is a cherished belief that the mass media is a forum where sincerely held views can be freely aired and debated. I am convinced that as far as the subject of evolution is concerned, the real truth is quite otherwise. There is however a glimmer of hope in three other countries.

Holland

The only European country where anti-evolution programmes are transmitted is Holland, where the Evangelical Broadcast group (Evangelische Omroep) has made a series of programmes on this theme.

Canada

A regular television programme entitled "Crossroads" is made at 100 Huntley Street, Toronto. It deals with Biblical exposition and the subjects of creation and evolution and is syndicated across all of Canada and part of the United States.

America

In America, various organisations have been publicising the evidence against evolution. One such body is the Institute of Creation Research which holds lectures and has travelling speakers debating the subject (when they can find opponents willing to accept the challenge). Another organisation is the Creation Research Society which publishes a scientific Quarterly. Due to the continual pressure from such groups, at last the mass media is beginning to acknowledge their existence.

England

The situation in this country, however, is quite the opposite, for the ban is almost total. A few articles have appeared referring to the weakness of the theory, but two aspects are most noticeable. The first is that, when the growing pressure of the Creationist Movement is referred to, there is little reference to the right or wrongs of their case, but it is used to urge the evolutionists to put their house in order and to present their case more carefully. The second point is one that I have already mentioned, i. e that there is never any reference to any important scientific *fact* which flatly contradicts the theory. To do so would bring the subject out of the sphere of metaphysics and philosophy, where hothouse theories can abound, into the cold light of day, when the inadequate clothing of facts around the theory would be clearly seen.

A Challenge

I will be prepared to admit that my charge of suppression by the media is unwarranted when I see a programme or article in which those who oppose evolution are given adequate opportunity (without editing!) to present the scientific facts which clearly support their case. Until then, I would maintain that my charge is proven. To those who consider this charge to be a gross exaggeration, I would suggest that they approach any of the organisations to which I refer and request that the whole subject of evolution be freely debated. I guarantee they will receive a refusal, which will be very polite — but a refusal none the less.

Reluctance to debate

It must not be thought that this reluctance to present a fair debate is a recent phenomena. The famous British Association meeting in

1860 can hardly be classed as a real debate, for as we have seen, there is no record of the arguments provided by either Wilberforce or Huxley and neither was there any voting on the outcome, which was seemingly far from being the 'victory' for evolution which its proponents claim. Indeed, Huxley's eagerness to debate appears to have consisted of lectures in support of evolution and a harangue against creationism. This deliberate avoidance of debate seems to have had a long history for Darwin, in his autobiography, mentioned some advice which Lyell gave him apparently many years before the publication of his *Origins*. Darwin said:

"I rejoice that I have avoided controversies, and this I owe to Lyell, who many years ago, in reference to my geological works, strongly advised me never to get entangled in a controversy, as it rarely did any good and caused a miserable loss of time and temper."[2 p89]

This consistent blockage of any publicity for the scientific evidence against evolution is met so frequently that the cause must lie not just at the level of academic intolerance but surely in the deeper areas of philosophy and even theology.

SECTION IV

PHILOSOPHICAL CRITICISMS

CHAPTER 29

THE PHILOSOPHY OF EVOLUTION

It is generally claimed that the theory of evolution is the only tenable scientific view of the Universe, as it is based upon a RATIONAL examination of the (NATURAL) MATERIAL world we live in. The theory sets out to show that the course of evolution was DETERMINED by purely mechanical considerations with no influences external to nature being required or even existing.

These three philosophies – Rationalism, Materialism (Naturalism) and Determinism – are the foundations upon which the thorough-going evolutionist depends. To clarify the views they express, I give the dictionary definitions as follows:

Materialist – "One who claims that nothing exists but matter and its movements and modifications and that consciousness and will are wholly due to material agency". (Similarly, Naturalist) i.e. nothing exists except matter.

Rationalist – "One who claims that reason is the foundations of certainty in knowledge".

Determinist – "One who claims that human action is not free but determined by motives regarded as external forces acting on the will". The thorough-going determinist would indeed take this further and say that the positions and combinations of every atom today were already determined by their relative positions following the primeval "Big Bang".

It will be immediately apparent that each of these views has one factor in common, i.e. the absence of any need for a God in order to explain the working of the present universe. It is this uniting factor which makes them so frequently common bedfellows, and it is in their company that evolution thrives. For this reason I make no apology for extending the scope of this section beyond that of the theory of evolution alone, for we must look behind the theory to discover its true source.

In this section, we will be examining the inherent weaknesses of the three views, and throughout I will use the names almost interchangeably, using, say, Rationalist when dealing with the aspects of logical argument, Materialist when considering the total universe, etc.

The whole subject of the philosophical basis of evolution is of considerable importance, but the arguments can sometimes appear to be involved. What follows is therefore a very simplified summary of the problem, a grasp of which, whilst not essential, will be well repaid. For those who wish to study the subject of the basis of scientific theories further, the works of Karl Popper are recommended [55].

THE EVOLUTIONARY ACCOUNT

If asked to give an account of how the present universe comes to exist, the Materialist would probably give the following historical outline.

"Originally there was a dense concentration of matter which exploded (The Big Bang). As this material moved through the surrounding space, it gradually condensed into stars. These then collected into groups to form galaxies. Some of these stars (suns) developed planets which circled them, and on some of them conditions were such that life could arise."

"One of them was this planet and, as it cooled, water condensed out to make oceans and, in them, various chemicals formed themselves by chance into complex molecules. Different molecules combined in such a way that they formed a primitive type of life which could duplicate itself. Further combinations resulted in cells which in turn gave rise to creatures. These creatures, in their quest for food, adapted to various conditions and eventually plants and animals appeared, the 'highest' of the latter being man. The whole process took millions of years, but as evolution is continuously developing, some superior form of man will eventually appear. However, there will be many 'hardships' as evolution is a random process involving trial and error, with only those who are fitted to the environment tending to survive".

Such a statement I would suggest gives the essence of the materialist-evolutionist survey of what might be called the Drama of the Universe.

In such a scheme, there is no need for any God (except for the unexplained existence of matter before the "Big Bang") and I will be dealing with this approach first, considering later those variations which believe either that it was due to a "Life-Force" or that God controlled the whole process of evolution (Theistic Evolution).

The rationalist's argument

The Rationalist would claim that he reached such a cosmological view by a series of logical steps along the following lines:

1) Rationalism is the only correct method by which man can discover Real Truth, as it is based entirely upon Reason and is free of presuppositions.

2) The universe (and life on this planet) is examined *as it is*, and numerous facts are gathered. From this he arrives at the idea that life gradually developed out of matter, and proposes the Theory of Evolution. On the basis of this theory, he examines many other facts, involving the whole universe, and claims that they (almost invariably) support the theory.

3) He therefore contends that his view is the only correct one as it is based upon reason, and is confirmed by the facts he investigates and he has therefore reached Real Truth. Consequently, all other methods of "knowing" – Intuition, Revelation (from God) and the occurrence of miracles – are classed as *un*-reasonable (i.e. unscientific) and can be accordingly dismissed and disparaged.

Thus, the Rationalist's belief in the 'scientific' method determines who can and who cannot be members of the scientific community. Those who do not accept this approach may be subjected to scientific 'ostracism'.

The rationalist's claim rebutted

As it is presented, the Rationalist's claim to be the *only* correct method of attaining a knowledge of Real Truth appears convincing. However, as with many bold claims, close inspection makes them far less credible. Firstly we will examine the nature of scientific thought, then in chapter 30 we will highlight the two great flaws in the Rationalist's approach, and finally in chapter 31 show that evolution is not a true science.

THE NATURE OF "SCIENTIFIC" THINKING
1. AREAS OF KNOWLEDGE

Our understanding of any sphere of knowledge relies upon many other aspects of knowledge. Thus a study of spatial dimensions depends upon the acceptance of the laws of mathematics. Similarly, a study of Biology depends upon the laws of Physics and Chemistry. In order to illustrate just how complex the whole subject of "knowing" is, I set out below the fifteen spheres of aspects of reality. Whilst they are a hierarchy, nevertheless they penetrate, and are penetrated by, all of the other spheres in varying degrees.

SUCCESSION OF SPHERES	SCIENCE
15. Pistical (faith)	Theology
14. Ethical (love)	Ethics
13. Juridical (judgment)	Jurisprudence
12. Aesthetic (harmony)	Aesthetics
11. Economic (saving)	Economics
10. Social (social intercourse)	Sociology
9. Linguistic (symbolical meaning)	Philology, Semantics
8. Historical (cultural development)	History
7. Analytical (thought)	Logic
6. Psychical (feeling)	Psychology
5. Biotic (life)	Biology
4. Physical (energy)	Physics, chemistry
3. Kinematic (motion)	Mechanics
2. Spatial (space)	Mathematics
1. Arithmetical (number)	Mathematics

It is only when we study such a list that we begin to realise how much is involved when we say we "know" a particular fact.

2) SELECTION OF EVIDENCE

In order to make an exhaustive study of a subject, the scientist would find himself absolutely overwhelmed by the total number of facts which could possibly be relevant. For example, imagine the situation of a scientist who wishes to examine the relationship between the extension of a spring and the load it carries. In considering *all* possible factors which might affect the result he may include –

a) The weight of the spring
b) The air temperature
c) The air pressure
d) The humidity
e) The amount of cosmic radiation
f) The accuracy of his instruments
g) His ability to measure and record correctly
... etc. etc. The list would be endless.

Obviously, the vast majority of such items he would dismiss from consideration. But we must note carefully two things he has done. Firstly, he can only reject certain factors *if he already has a conception of what the relationship is* — that is, he has a theory which he wishes to test. Secondly, he has ignored certain factors which *in his opinion* will not affect the results. Nevertheless, there is always the possibility that they *may*, but by excluding them he has made a value judgement whether he realizes it or not.

Occasionally, it is an ignored factor which is later found to play a vital role in a particular experiment or theory. A major example of this occurred towards the end of the last century. The basic laws governing Chemistry, Physics and the major sciences had been discovered, and it was confidently felt that it only remained for the precise nature of atomic particles to be clarified for man to have found all the existing physical laws of the Universe. It was the experiments of the atomic physicists, however, which revealed, amongst other things, that atoms did not obey Newton's Laws of motion. A whole new sphere of investigation was thrown open, and even today many basic questions remain unanswered.

3) INTUITION

Whilst the scientist may claim to examine facts in isolation, nevertheless much of our knowledge is tacit and intuitive. Popper points out that:

"...every organism has inborn reactions or responses...These responses we may describe as 'expectations', without implying that these 'expectations' are conscious....In view of the close relation between expectation and knowledge we may even speak in quite a reasonable sense of 'inborn knowledge'. This inborn 'knowledge' is not, however, *valid a priori*; an inborn expectation, no matter how strong and specific, may be mistaken....We are thus born with expectations....One of the most important of these expectations is the expectation of finding a regularity. It is connected with an inborn propensity to look out for regularities, or with a *need* to *find* regularities..." [55 p47].

Thus, in our examination of a varied array of facts, we (and scientists) subconsciously expect to find (have theories regarding) certain regularities (laws) between them. This subject of how scientists approach evidence with already formulated theories, we will now consider.

4) THE "DISCOVERY" OF SCIENTIFIC LAWS

It is generally thought that a scientist first collects a mass of facts, and then by carefully inspecting them he sees a general pattern. He then tentatively proposes a theory and, by making further experiments, finally produces a "law". It was this approach which Darwin claimed when he said: "I worked on truly Baconian principles, and without any theory collected facts on a wholesale scale..."

A careful examination by Popper of how scientists actually make their discoveries, however, shows that the actual facts may be very few, whilst the idea is more like a flash of light or inspiration.

Independent confirmation of Popper's views was provided in a very perceptive article by Michael Polanyi, entitled "From Copernicus to Einstein" [31]. He begins by saying:

"In the Ptolemaic system, and in the cosmogony of the Bible, man was assigned a central position in the universe from which he was ousted by Copernicus. Ever since, writers eager to drive the lesson home have urged us, resolutely and repeatedly, to abandon all sentimental egoism, and to see ourselves objectively in the true perspective of time and space. What precisely does this mean?"

The main point of his article is that scientists are not as objective as is commonly supposed. He claims that they indulge in the very subjective practice of intellectual satisfaction in proposing and testing elegant theories and not with a purely rational and objective approach.

He provides two well known examples, the first being the discovery of the laws of planetary motion.

It is obvious that the sun, moon and planets appear to circle the earth. There are however some anomalies. Copernicus and Kepler both deliberatly returned to the Pythagorean quest for harmonious numbers and geometrical excellence. It was to satisfy this aim that the proposal of the planets moving around the sun was made.

Furthermore, in reading Kepler's works it becomes obvious that he is really a sun worshipper, when he makes such comments as "Of which sort of vision is in the sun, what are its eyes, what other impulse it has...for judging the harmonies of the motions.... In the sun there dwells an intellect simple, intellectual fire or mind, whatever it may be, the fountain of all harmony." He went so far as to write down the tune of each planet in musical notation and indeed it is clear that to him astronomical discovery was an ecstatic communion, as witnessed by some flamboyant writing.

The second example Polanyi gives deals with Einstein's discovery of the Theory of Relativity. The Michelson-Morley experiment carried out in 1887 was to determine the speed of the aether as the earth travelled through space and the effect this had upon the speed of light. Much to the surprise of the physicists, the results were said to be virtually zero.

It is frequently claimed by books dealing with the history of science that it was in order to explain this result that Einstein formulated his Theory of Relativity. As is so often the case, the truth is quite otherwise. When he was only sixteen years of age, Einstein, in the course of theoretical speculations regarding motion and light, revealed a paradox which the known laws could not explain. It was to deal with this that he finally produced his special Theory of Relativity. Einstein confirmed that it was not due to any experimental results, for he wrote to Polanyi that "the Michelson-Morley experiment had a negligible effect on the discovery of relativity".

Polanyi continues by revealing the "almost ludicrous part of the story to be told". The actual results of the Michelson-Morley experiment showed a small positive "aether-drift" of 8-9 Km per second, which was verified by numerous highly sensitive experiments. Polanyi continues:

"The layman, taught to revere scientists for their absolute respect for the observed facts, and for the judiciously detached and purely provisional manner in which they hold scientific theories (always ready to abandon a theory at the sight of any contradictory evidence), might well have thought that, at Miller's announcement of this overwhelming evidence of "a positive effect" in his presidential address to the American Physical Society on December 29th, 1925, his audience would have instantly abandoned the theory of relativity. Or, at the very least, that scientists - wont to look down from the pinnacle of their intellectual humility upon the rest of dogmatic mankind - might suspend judgement in this matter until Miller's results could be accounted for without impairing the theory of relativity. But no; by this time they had so well closed their minds to any suggestion which threatened the new rationality achieved by Einstein's world picture that it was almost impossible for them to think again in different terms. Little attention was paid to the experiments; the evidence being set aside, in the hope that it would one day turn out to be wrong."

Polanyi's ironic quip of scientists being "always ready to abandon a theory at the sight of any contradictory evidence" exactly echoes Darwin's ridiculous claim that "I have steadily endevoured to keep my mind free so as to give up any hypothesis however much beloved (and I cannot resist forming one on every subject), as soon as facts are shown to be opposed to it."[2p103]

5) THE PROBLEM OF ORIGINS

It will be obvious that the main business of science is to discover (and use) the various laws governing the natural world, which it does by examination of the present state of the universe. What it cannot do is "prove" in a similar fashion how the universe first originated, for this is a problem of quite a different order and beyond its scope. It is the difference between asking "How does it work?" and "How did it come to exist?"

This does not prevent scientists putting forward a hypothesis on how the universe originated, but whether it is acceptable will depend upon how much corroborating evidence is produced, and whether an alternative theory may be better supported.

This is indeed a fundamental aspect, and one which we will show is bound up with one's philosophical point of view, when we consider the second flaw in the Rationalist's argument later.

SUMMARY

In the five sections above, I have only briefly contradicted what might be called the "popular" view of science, showing that scientists progress by intuition and by limiting their area of investigation. This they should surely admit rather than claim that their approach is the only source of true knowledge.

Whilst the more humble scientists are fully prepared to make such an admission, nevertheless the popular scientific journalists rarely refer to this aspect. They invariably imply that there is no true knowledge outside that which has been "proved by science".

So successful is this image that, for the general public, the final authority they can quote is "Science proves..." Careful questioning, however, often reveals that such "authority" is a statement made on behalf of "science" by a popular personality in a television broadcast yesterday evening!

T.H. Huxley once wrote:

"Sit down before fact as a little child, be prepared to give up every preconceived notion, follow humbly wherever and to whatever abysses nature leads, or you shall learn nothing" [13p316].

Such an initial approach, as I have agreed, is correct, for there is a limit to the number of facts our minds can try to correlate. But implied in his statement (by his silence regarding the limitations of this approach) is the idea that through this method, and this method alone, lies the path to all knowledge of Real Truth. I would compare his call to be rather like asking a student to examine a small electrical component, without telling him that it is part of a complex computer.

It is of course perfectly acceptable that scientists should don glasses when making examinations. But we must reject both the claim that only *he* sees the truth, and furthermore, that he is not wearing spectacles!

C.S. Lewis has set out the rise of scientific arrogance in its historical setting, for he wrote:

"... since the Sixteenth Century, when Science was born, the minds of men have been increasingly turned outward, to know Nature and to master her. They have been increasingly engaged on those specialised inquiries for which truncated throught is the correct method. It is therefore not in the least astonishing that they should have forgotten the evidence for the Supernatural. The deeply ingrained habit of truncated thought - what we call the "scientific" habit of mind - was indeed certain to lead to Naturalism, unless this tendency were continually corrected from some other source. But no other source was at hand, for during the same period men of science were coming to be metaphysically and theologically uneducated."[32p46]

CHAPTER 30

SELF CONTRADICTIONS

The Rationalist/Materialist's claim that he is guided by logical reasoning deduced from examined facts in arriving at a "true" concept of the Universe suffers from two fundamental flaws, which strike at the very heart of this approach.

1) FIRST OBJECTION :
MATERIALISM VERSUS RATIONALISM

It is the Materialist who claims that the Universe is nothing but a chance concourse of atoms in which certain chemicals combined in increasingly complex ways eventually to produce Man. Man then looks back over the whole of time and grasps within his mind the whole sweep of the drama of how the Universe developed, and how he himself arrived.

There are two objections to this claim, both dealing with the same aspect, but from slightly different points of view.

A) *Randomness versus true knowledge*

It is not possible that a random process, no matter how complex, could ever obtain any insight of Real Truth. Thus, if our minds are only a complex interaction between atoms, how can we be certain that the "thoughts" they give rise to are *true* reflections about the *real* world?

Darwin saw the importance of this problem, for he admitted:

"But then arises the doubt, can the mind of man, which has, as I fully believe, been developed from a mind as low as that possessed by the lowest animals, be trusted when it draws such grand conclusions? I cannot pretend to throw the least light on such abstruse problems. The mystery of the beginning of all things is insoluble by us; and I for one must be content to remain an Agnostic" [2p313].

Professor Haldane similarly concluded:

"If my mental processes are determined wholly by the motions of atoms in my brain, I have no reason to suppose that my beliefs are true...and hence I have no reason to believe my brain is composed of atoms".[72p209]

One argument put forward to avoid this dilemma is to say that we can reach true knowledge by inferences; that is, that whilst we cannot see how sub-rational behaviour can achieve true insight, the Rationalist would claim it has, for natural selection preserves useful behaviour, and if useful, it must be true. But as C.S. Lewis shows [32p25], this last statement (if useful, then true) is itself only another inference. If the value of our process of reasoning is being critically examined, then we cannot establish its truth by the process of reasoning itself!

Similarly, it could be argued that, if natural selection enables man to survive, then his thought processes are also selected for their survival value alone, and *not* for whether they are really true or not.

Thus from a purely Materialistic viewpoint, we can never prove that man can grasp reality by reasoning.

B) *Nature versus Rationalism*

Any process which observes another process must itself be outside the action it is observing. For example, a scientist can study the laws of gases, i.e. their volume, temperature, pressure, etc. On a different level, a specialist can study *how* a scientist goes about examining and theorising on scientific subjects, as Karl Popper has done. Yet again, the scientist himself may go over in his mind the method by which he arrived at a certain theory. In all these examples, the objects (or method of thinking) being considered are 'outside' of and quite separate from the thought processes of the person whilst he is considering *how* a theory was formulated.

The Naturalist, however, will say that there is nothing outside Nature (which would include his own thought processes), yet by thinking about Nature, he claims he is able to stand outside Nature and contemplate the whole sweep of history. His mind, which is composed entirely of atoms (and therefore part of the natural world), he claims can at the same time have thoughts which are outside the natural world, which is logically inconsistent.
C.S. Lewis points out that:
"the knowledge of a thing is not one of the thing's parts. In this
sense something beyond Nature operates whenever we reason."
Thus again we meet the same dilemma as before in a different form, i.e. a pure Rationalist/Naturalist approach can provide neither proof that rational thinking is really true, nor provide a standpoint from which the whole of Nature can be surveyed.

2) SECOND OBJECTION :
THE UNIFORMITARIAN ASSUMPTION

The Rationalist will only consider present states - what cannot be examined and tested is not accepted. As we do not experience miracles, such as recorded in the Bible, in our present age, he therefore claims that all events confirm the Uniformitarian position.

But this assumes that the future (and past) is like the present, i.e. it assumes Uniformitarianism is true. Therefore it is a basic assumption and is not open to proof.

Similarly, if all experience is absolutely Uniform, then reports of miracles must be false. But we can only 'know' they are false if we 'know' that miracles have never occurred. Thus, as we have shown above, the Rationalist is caught once again in a circular argument of his own making. If man limits his investigations to the few short years he exists on this earth, then he is bound to be a Uniformitarian. Indeed, just as man has tended towards an evolutionary view of life, so has he also inclined to a Uniformitarian view.

As always, C. S. Lewis neatly summed up the position when he wrote:

"It is no use going to the texts [of the Bible] until we have some idea about the possibility or probability of the miraculous. Those who assume that miracles cannot happen are merely wasting their time by looking into the texts: we know in advance what results they will find for they have begun by begging the question" [32p8].

It is for this reason that we cannot leave the question of origins to the "unbiased scientist" – the religious issue is too fundamental.

To conclude, there is a further aspect of the Uniformitarian assumption which is worth pondering upon. It claims that "the present is the key to the past". This is a complete reversal of the Christian faith which says that the past (man's rebellion against God) is the key to the present (the problems of the world today)!

CHAPTER 31

IS EVOLUTION SCIENTIFIC?

The scientific aura which surrounds evolution is at long last being seriously questioned. Even in the British Natural History Museum's publication "Evolution" [22], the author, Colin Patterson, poses this question and makes the tacit admission that the theory of evolution is not scientific, according to the definition of a true science by Karl Popper [p149].

Popper has subjected the scientific method to a critical examination and has come to some remarkable conclusions. His interest in this subject stems from the time when, as a committed Communist for a short period, he was shocked when some colleagues were killed in a clash with the police. He thereupon subjected the Marxist claim to be based on the "scientific laws of historical social development". After several years of study, he saw clearly how false this claim was; its use of the label 'scientific' (and its implication of inevitable truth) was deceptive in the extreme. As the Marxist theory is founded upon the 'progress' of evolution, he showed that this also was unscientific. This led him to examine the whole quesion of what constitutes a true science and what is only a pseudo-science?

Summarising Popper's view briefly, a true science is one in which experiments can be devised which could refute the theory under consideration. In a pseudo-science, no experiment which would finally refute a theory can be made (usually because some other explanation for non-compatible results can be found).

For example, a theory regarding the relationship between heat and temperature can be tested in any laboratory at any time and can therefore be classed as scientific. On the other hand, evolution has only taken place once in the past and therefore no 'experiment' of evolution can be carried out. All that can be done is to study such subjects as the fossils in the rocks and offer an interpretation. Even if experiments on animals were to show that completely new species could be made (which they have not yet achieved), this would only indicate that evolution *could* have occurred, not that it *had*. Thus, by Popper's definition, evolution certainly cannot be classed as a true scientific theory.

Popper also refers to the different attitudes of the true scientist and the pseudo-scientist. Examples of the latter would be astrologers and psychoanalysts. The true scientist will devise as severe an experiment as he can in order to test whether the theory can be refuted, and if it is, he will reconsider the theory.

The pseudo-scientist on the other hand will only look for evidence which will confirm his ideas, and should he be faced with contrary evidence, will simply provide secondary theories in order to explain them away. With this in mind, even a cursory reading of *The Origin of Species* will show that the author is a past master of the art of 'explaining away' unacceptable facts.

A further aspect is that pseudo-scientific theories, of which evolution is a prime example, do not usually make exact predictions of what will occur in any particular situation. Evolution can only look at an organism and then provide a reason for some of its distinguishing characteristics. It cannot *predict* how an animal will develop under particular environmental conditions. This is but one further example of how evolution is beyond proof by the usual methods of scientific verification.

It is at this point that I would refer to one aspect which Popper raises, which perhaps few people realise. This is that no experiment can ever *prove* a scientific theory, it can only *disprove it.* This may appear surprising, but the logic behind this is as follows. Each experiment may support or disprove a particular theory. No matter how many experiments are carried out which support the theory; it may be that the one (or more) experiment which *would* disprove it has yet to be made. Every experiment which complies with the theory only makes it more probable that the theory is correct. Thus the scientist can only claim that *so far* his theory has not yet been disproved. No single experiment (or a series) can ever finally 'prove' a theory in the strictest sense of the word - it only has the potential to disprove it.

SUMMARY

So far, then, we have tried to show that the Rationalist/Evolutionist approach to scientific investigations is *not* the monolithic, unshakable foundation of all knowledge which it claims for itself, but is actually bound by its self-imposed blinkers, is founded upon circular arguments which destroy its logical basis and is ultimately unprovable. In short, it possesses all the same innate failings and self deceptions to which all purely human endeavours are subject.

CONCEPTUAL FRAMEWORKS

It is here that the term "Conceptual Framework" should be introduced. This is best described as a generalised Theory (or Axiom) about a number of separate theories. Put more simply, it is as its name implies- a framework in which the relationships between various theories can be unified and thereby conceived. Examples of different conceptual frameworks would be atomic, thermodynamic and quantum theory in Physics and Chemistry, and Mendelism and

Cell Theory in Biology. A scientist who supports a particular view will bring that C.F. to bear in any theory he may formulate.

To give an example: a scientist may study the laws of heat and temperature, and after various experiments may theorise that the temperature of a body is proportional to the heat it absorbs. However, he then heats water which eventually turns into steam, and he finds that the heat in the steam plus that in the water is less than the total heat he put into the water. Does he abandon his theory? No- because he has the Conceptual Framework regarding heat which says that energy is not destroyed or 'lost' i.e. The First Law of Thermodynamics. He therefore investigates further and proposes that the 'missing' heat be labelled "the latent (hidden) heat of steam".

This is only an oversimplified example to illustrate how C.F.s override particular theories and non-complying results. If conflicting results do arise which cannot be explained, then they are either put to one side (or ignored), in the hope that a theory may later arise which will provide an explanation.

The reason for describing C.F.s in some detail is to point out that they are not refuted by individual contrary facts. Such facts may disprove a particular theory, but the Conceptual Framework itself will remain inviolate. It is only when a number of theories, upon which it rests, are refuted, will the truth of the C.F. be called into question.

Furthermore, linking this to the subject of sciences versus pseudo-sciences, a C.F. can only be considered as scientific if the theories it covers are themselves scientific (i.e. capable of being disproven).

From this it is obvious that Darwin's theory of evolution is not just a theory but a Conceptual Framework and, moreover, a pseudo-scientific one, resting as it does on pseudo-scientific theories.

That it is a Conceptual Framework is evident from the large number of theories it has spawned which attempt to explain how all living organisms gradually evolved to be as we know them today. It has an immense capability of explaining existing forms and systems even beyond the sphere of biology, for it is now used to interpret 'developments' in areas of study as different as astronomy and sociology. More important, in the fields of economics and politics, evolution forms the analytical foundation and is the basis of the Marxist philosophy.

In order to show that evolution is a pseudo-scientific Conceptual Framework, I will give just one example (out of many that could be given) of how it overrides the evidence which contradicts it. I have deliberately selected one from astronomy in order to show how wide ranging and influential the theory is when applied to areas of knowledge far removed from the biological field which fostered its growth.

The age of comets

There are a number of comets which have elliptical orbits around the sun, some of which return every few hundred years whilst others have return periods of thousands of years. Each time a comet passes close to the sun, it loses volatile matter which is vapourised off by the sun's radiation, and it is this "lost" matter which forms the tail of the comet. Some comets have short return periods of only a few hundred years - Halley's comet in fact returns every 76 years. It can be shown by simple calculation that these short period comets would have disappeared within a matter of 10,000 years or so. The most obvious deduction from this is that they could not have been in existence for more than, say, 100,000 years.

Now evolutionists claim that the earth was formed 4,500 million years ago, whilst the age of the Universe (from the time of the Big Bang) is thought to be in the order of 10-20 billion years. What happens when presented with the obvious evidence that the planetary system at least is vastly much younger? Do they acknowledge the results as contradicting their views. Of course not. They have to recourse to what we might call an explanatory theory, i.e. a theory which is deliberately invented simply to enable them to accommodate these 'awkward' scientific facts within the Evolution Conceptual Framework. Such a theory was provided by Oort, who suggested that there was a cloud of numerous long term comets orbiting the planetary system effectively in 'cold storage' in outer space, which are occasionally disturbed from their normal orbits by passing stars. Some would be perturbed by Jupiters gravitational field and move in the direction of the sun, which would then capture them within its gravitational field. They would then become one of the unending series of short term comets necessary to maintain the evolutionary long term framework.

Now there is not the slightest scrap of evidence which supports Oort's theory for there is no evidence of a vast number of very long term comets and Jupiters gravitational field would not be sufficient to provide the number of short term comets we have today [69]. Yet it is the stock answer given by astronomers when challenged with the evidence. Why should men, who claim to be logical and objective in their field of study, have recourse to such speculative theories, when the clear evidence points in a quite different direction? The answer will be obvious from what I have set out previously. Theories such as Oort's *have* to be accepted simply because an important cornerstone of the evolution theory – in this case the vast time spans it requires – must be preserved against all evidence to the contrary.

In such a fashion is evolution propped up by a number of 'explanatory' theories which have little or no evidence in their support. Many similar examples could be given such as —

1) Goldschmidt's 'hopeful monster' theory to explain the way in which mutations might provide a new species.

2) "Punctuated Equilibrium Theory" i.e. that species have long periods with little or no change between short periods of rapid development and change due to local environmental conditions. This is only proposed in order to explain the large gaps in the geological fossil record.

As I have said, no Conceptual Framework is questioned until there is a considerable amount of doubt surrounding its supportive theories. Only then is the Conceptual Framework itself reconsidered. The evolution C.F., however, is a notable exception. No matter how many facts against it may be presented, and explanatory theories invented to accommodate them, the central evolution C.F. is never questioned. Indeed, it has been likened to a house built on props, every one of which has been removed but the house still stands !

The simple reason for this state of affairs is that for those who dogmatically hold to evolution, the abandonment of the theory would involve a cataclysmic reassessment of their personal philosophy of life. Such a process would be so painful that they avoid it at all costs.

Willful blindness

We must at this point question the way in which scientists interpret the evidence before them. There are countless areas where the facts are plainly pointing to the existence of a Designer and flatly contradict the possibility of evolution by 'chance'. In astronomy, we have already shown how the evidence for the young age of comets is deliberately circumvented. Similarly there are numerous 'anomalies' in the planetary system (see appendix V) which ought to indicate to astronomers that the planets were created in (virtually) the same relationships and material as we see them today. Yet how many of them accept the simple inference which should be drawn ?

This situation is repeated in all those fields where the natural world is studied for they all display an incredibly complex organisation. In physics, experiments have shown that even in inert matter, there are the innumerable fundamental particles and the mathematical relationships of atomic structures and materials. In biology, there is the complexity of many forms of life and the amazing amount of information which is stored in the genes and chromosomes. In agriculture, the planting of minute seeds can give rise to an enormous tree or a beautiful flower. Doctors and Nurses have to study the complex sequence of biochemical and cellular changes which have to take place in the devopment and birth of a baby. There are many other areas where a long list of such facts could be drawn up.

In any study of nature, whenever the evidence is investigated in any detail, there comes a point when the facts simply *shout* at us that they are the result of a deliberate plan by a creator God. This was accepted for many generations by Christians as an acknowledged basic fact of life. Yet men have always deliberately ignored them and the lesson that they carry - more so today than ever before. Indeed, one is forced to the conclusion that those who consistently ignore the evidence of superb design are being not just blind but *wilfully* blind. This is clearly predicted in the Bible which warns us that men will "suppress the truth" for "what may be known about God is plain to them for God has made it plain to them....so that men are without any excuse." [Romans 1 v 19]. So striking are such passages in the Bible which refer to the deliberate blindness of mankind that is particularly evident today that these have been brought together in Appendix VIII together with other relevant topics.

Alternative Frameworks

In scientific investigations, there are at times two or more Conceptual Frameworks which provide alternative views of various phenomena. As more evidence is gathered, it often becomes clearer which of the alternatives provides the better explanations, and, after a time, the others are abandoned.

There is, however, an important consideration. When the evidence against a Conceptual Framework begins to mount up, then there will come a time when it becomes obvious that the Conceptual Framework provides a very unsatisfactory basis. However, *it will not be abandoned if there is no acceptable alternative Conceptual Framework to turn to*. To scientists, the discovery of laws which connect various phenomena is an important consideration, and this being so, then *any* Framework, even one which has many faults, is better than none at all.

Evolution is a prime example of a Framework which is clung to despite the great array of well documented facts against it. As I have said, these facts rarely reach the general public, but ample evidence against the theory is provided by the many creationist books on the subject. It is not the purpose of this book to set out such evidence in detail, but a synopsis of some of the more important facts are given in Appendices III-V.

Despite such evidence, evolution is still retained, only in this case, not because there is no alternative to go to, but as the alternative Conceptual Framework (Special Creation) is simply unacceptable to those who hold to Evolution, regardless of the facts against it.

Confirmation of this was given by Professor D. M. S. Watson in an article in Nature in 1929 (v124 p233). In a statement which is notable for its frank honesty, he admitted:

"...the Theory of Evolution itself [is] a theory universally accepted not because it can be proved by logically coherent evidence to be true, but because the only alternative is special creation, which is clearly incredible."

CHAPTER 32

ALTERNATIVE VIEWS

So far I have considered only the concepts of Evolution and Special Creation. There are, however, other views which might be considered as variations or intermediate positions between them.

1. NON-EVOLUTION

Amongst many who have studied the evidence for and against evolution, both scientists and laymen, a number are willing to admit that it is very unlikely that evolution took place. At this point, however, they stop, for again the idea of a God who can create instantly from nothing is either too vast for them to comprehend, or simply does not accord with our scientific age, when men restlessly seek to find a rational cause for every event.

They are thus effectively in a philosophical limbo, where the facts reject one viewpoint and their limited theology baulks at the other. In such situations, the escape route often used is simply to say: "It doesn't really matter all that much, does it? We find ourselves in this particular world and we have to live as best we can without getting too involved in theoretical questions which have no bearing on what we do in our daily lives". Such an attitude is held by those who do not wish to pursue the problem to its logical conclusion for they know instinctively that the implications are enormous if Creation by an all-powerfull God were shown to be true.

2. LIFE-FORCE

As Evolution effectively enabled man to get rid of God, it was not long before an alternative power was proposed to fill the resulting vacuum. This was the claim that there was in Evolution itself a creating force which expressed itself in the 'purposive' strivings of the various organisms to compete against each other, thus producing more complex and efficient types.

One of the main exponents of the Life-Force (or Creative Evolution) view was Bergson. More recently, the writings of Teilhard de Chardin are an attempt to give a Christian gloss to the same concept, by claiming that the ultimate end of evolutionary development was for the Universe to progress both materially and spiritually until it achieved a mystical union with God - the Omega point. Although Teilhard's views still receive considerable publicity from some organisations, they are now much less discussed than they were in their popular heyday. As far as the activities of Teilhard de Chardin are concerned, I have shown in my first book the very important part he played in the Piltdown hoax and in the 'discovery' in various countries of many of the virtually fraudulent series of fossils which are purported to 'prove' man's ape ancestry.[46]

As C.S. Lewis points out, if the Life-Force has a mind, then it is really a God in a different form, i.e. it is a Religious view. If it has not, then one cannot claim that it "strives" or has "purpose". As Lewis comments:

"One reason why many people find Creative Evolution so attractive is that it gives one much of the emotional comfort of believing in God and none of the less pleasant consequences. When you are feeling fit and the sun is shining and you do not want to believe that the whole universe is a mere mechanical dance of atoms, it is nice to be able to think of this great mysterious Force rolling on through the centuries and carrying you on its crest. If, on the other hand, you want to do something rather shabby, the Life-Force, being only a blind force, with no morals and no mind, will never interfere with you like that troublesome God we learned about when we were children. The Life-Force is a sort of tame God. You can switch it on when you want, but it will not bother you. All the thrills of religion and none of the cost. Is the Life-Force the greatest achievement of wishful thinking the world has yet seen?" [30p34].

3. THEISTIC EVOLUTION.

By far the most common view of those who believe that a God exists is to say that the whole of the history of evolution took place under the control of God. To those who hold this view, I would make the following points:

A) Unscientific

There are many thousands of true Christians who do not accept the Genesis account of Creation mainly because they are firmly convinced that the whole weight of scientific evidence is against it. Every book obtained from a library on geology or biology constantly asserts that evolution is a well proven fact. They therefore understandably look upon Creationists as deliberately avoiding the truth. In illustration of this I would quote the case of Prof. Rendle-Short, a well known Christian writer. Perturbed about the the conflict between evolution and creation, he studied Geology and obtained a degree in the subject. As a result it would seem that he reluctantly agreed with theistic evolution. Here was a brilliant man who conscientiously studied geology yet failed to find any support for the Bible. Why should this be so ?

For the answer, we must go right back to the work of Lyell in propogating the Uniformitarian theory. As we have shown, all the universities subsequently adopted the theory and as a result, *every textbook for generations to come would be written entirely from the point of view of the Uniformitarian Theory* and only evidence which supported it would be published. Take, for example, the well known 'horse series' of fossils which purport to show the development of the horses hoof. This has since been denounced by evolutionists as "a deceitful delusion" (see Appendix VII) and consists of fossils artificially arranged in a sequence that is virtually fraudulent. Yet it is evidence such as this which has been used as a propaganda weapon for evolution for many generations. Similarly today, many accept the "millions of years" provided by radiometric dating as "scientifically proven". Yet again, the large number of contradictory results and the unproveable assumptions on which they are based render the whole method totally unreliable.

It is therefore little wonder that those who consult only the 'standard' works or popular books are unwittingly indoctrinated with uniformitarian concepts. It is only comparatively recently that creationist scientists began to re-examine the strata and give publicity to a far more convincing reinterpretation in support of Catasrophism. As we have seen , Marxist evolutionist (e.g. S.J. Gould) are also providing a Catastrophist explanation. There are therefore a number of books available in Christian bookshops to which the enquirer can turn to in his search for truth. [Dr. Duane T. Gish's *Evolution — the fossils say NO* is particularly recommended.]

Similarly in the field of biology, much of what passes as 'clear proof' of evolution is found to be very insubstantial upon close inspection. I would strongly recommend anyone who has never seriously considered the evidence against evolution to purchase a well documented book on the subject and study it with as open a mind as possible. The result cannot be predicted, but I would suggest that at the very least, he would have to acknowledge that the theory was based upon foundations which are much shakier than he ever imagined. Many who accept Evolution do so simply because they have never studied the evidence against it. I would also state at this point that I have never yet found a single piece of evidence for evolution which, no matter how well it appears to be supported by the evidence, cannot be heavily criticised by a creationist scientist in that particular field.

B) Social Pressures

Many parents, as they see their children willingly adopting the fashions, habits and values of the group of young people with whom they mix, recognize how powerful are the psychological pressures of the group to conform to its accepted standards. When one is detached from a particular situation, it is far easier to recognise the various relationships and cross currents in such a group. Seeing this, parents may complain, "If only they would think for themselves instead of just conforming to the herd." Such comments invariably carry the suggestion that the parents are themselves free and independent of group pressure, which, in varying degrees, is simply not true of any of us.

That evolution is an "accepted scientific fact" in our present day society exerts a powerful influence upon us all to accept the statement as correct. Indeed, even to suggest the possibility that it may not have occurred is likely to raise a few eyebrows. On fundamental issues, to hold views which are radically different from those of our colleagues and friends is to invite a degree of ostracism which few are prepared to withstand, unless receiving support from some other source.

I would suggest that it is this pressure which, more than any other, prevents many, whether churchgoers or not, from accepting the Genesis account of Creation. This pressure is even greater upon those who work in scientific fields. I am quite convinced that there are many scientists who either profess to evolutionary views whilst remaining unconvinced or avoid the subject altogether, simply because to say that they do not accept evolution may cast a blight upon their career prospects or even jeopardise their postion. This may seem to be an exaggerated claim, but I am certain that this pressure is far more widespread than most people realize. To test this, I would ask all those who work in an area of science, "What would be the reaction of all your scientific colleagues if you were to asert firmly that evolution was not an acceptable scientific theory?"

Similarly, to test the strength of the conviction of any who hold to Theistic Evolution I would ask, "If *all* your friends were Creationists, would you still hold to your views?"

C) Concept of God

The concept that our Universe developed to its present state over many millions of years has become part of our educational and cultural inheritance. With such a background, the suggestion that God brought it into being in stages by instantaneous acts of creation over a period of six days comes as a distinct shock. There is an immediate resistance to the idea and the objection that "God does not work like that". Indeed, there is the lingering thought that God *cannot* create complex systems instantaneously, but needs to take time to plan and develop them.

Such a view is surely due to imposing our human limitations upon our concept of God. Because *we* have difficulty in executing complex projects, it is so easy to infer that God has to work in the same fashion.

Closely allied to this limited view is the similar failure to acknowledge the total powerfulness of the God to whom many pay lip-service. It is easy to conceive of a God who is loving and protective towards those who worship him, for we can relate that to human attributes. But a God who can create — and by inference annihilate — in an instant of time requires an enlargement of one's conceptions which some may baulk at. "Gentle Jesus, meek and mild" is an aspect of God which is willingly accepted. However, a God who *at the same time* possessed unlimited power to do whatever he wishes is a vastly different proposition, for it would have some very uncomfortable consequences. There are doubtless those who are christians in name only, for the God they worship is only pocket-sized, to be brought out only on religious occasions.

D) Support for Atheism

The claim that God controlled the whole process of evolution is a quite unproveable assumption and raises the suspicion that it is put forward simply to get God 'into the picture'.

In defending evolution from attack because of their religious views, theistic evolutionists are supporting a theory which is perfectly capable of being interpreted from a purely materialistic point of view. When a theistic evolutionist makes a statement or gives the results of experiments, the atheist has no difficulty in ignoring any reference to God — implied or explicit — and then incorporating the rest into his philisophy of life.

Sometimes theistic evolutionists accuse creationists of "doing the devil's work for him" by deflecting attention away from preaching the gospel of salvation by faith in the propitiatory death of Christ alone. In view of the assisstance which they give to the atheist, I would suggest that it is the Theistic Evolutionists who reconsider their own position in the light of such a charge.

This charge that creationists distract from the preaching of the gospel appears to be a valid criticism. However, we are all called to "give a reason for the hope that lies within us". Those who are involved in the creation/evolution debate can testify to the thought provoking response obtained from those who may never go to church. Indeed, it is encouraging to know that one American evangelist has said that he much prefers to go to university campuses which have been previously visited by creationist speakers for he finds that far more students are prepared to make a full personal commitment to the true Christian faith.

In summary, I would plead with all who accept Theistic Evolution to reconsider the whole subject, for I would contend that it is scientifically unacceptable, maintained by social pressure and dishonouring to God.

To clarify the position even further, I would pose the following question to theistic evolutionists. If God is not limited in power and *could* have created the world, if he has given man a record of what he did, and furthermore, if the scientific evidence does not contradict it, *what then prevents you from believing that it actually took place ?*

GENERAL SUMMARY

In this section we have examined the philosophical roots of the theory, shown its self contradicting errors of logic, revealed its pseudo-scientific status and criticized some theological variations.

All this would be important even if it was simply a scientific theory dealing solely with the mechanism by which life arose on this planet. However it is far more influential than a scientific theory for it has permeated the whole of Western society and is used as a framework in both interpreting the Universe and in justifying the course of events.

If the theory is false, it does not take much imagination to see the social repercussions this would have. Indeed, the implications of the theory are more important than the theory itself and to this we will now turn.

SECTION V

IMPLICATIONS AND APPLICATIONS

CHAPTER 33

MORAL REPERCUSSIONS

In his *Origins*, Darwin was well aware of the theological implications of 'his' theory, and was very careful to avoid this aspect. He dealt entirely with the animal and plant kingdoms, but he knew that once he had proposed that all life came from a single source, then there was the obvious inference that Man himself evolved entirely due to suitable environmental conditions, and that God (if he existed at all) was not an active agent in the process. Darwin wrote to a friend:

"With respect to man, I am very far from wishing to obtrude
my belief; but I thought it dishonest to quite conceal my opinion.
Of course it is open to every one to believe that man appeared by a
separate miracle, though I do not myself see the necessity or
probability" [3p263].

The reason for Darwin's restraint is interestingly given in the very next letter which appears in his *Life and Letters*. He is writing to Lyell and says:

"I shall be truly glad to read carefully any M.S. [manuscript]
on man, and give my opinion. *You used to caution me to be
cautious about man.* I suspect I shall have to return the caution a
hundredfold!" [3p264].

Both men were clearly aware of the size of the explosion which would result from the fuse they had lit.

Sedgwick was also well aware of the implications of the theory, for, having read the book, he wrote to Darwin, pointing out the important link between our morals, and our interpretation of the material world. He says:

"You have ignored this link; and, if I do not mistake your
meaning, you have done your best in one or two pregnant cases to
break it. Were it possible (which, thank God, it is not) to break it,
humanity, in my mind, would suffer a damage that might brutalize
it, and sink the human race into a lower grade of degradation than
any into which it has fallen since its written records tell us of its
history" [3p249].

Unfortunately, Sedgwick was wrong. The link *can* be broken, and man can interpret the Universe in any way he wishes. With Darwin's array of 'facts' to support him, he has done so, and his motivation has gradually become increasingly degraded; a trend to which there seems to be no limit.

This passage by Sedgwick is referred to by Clark, who has an excellent chapter where he traces the way in which the theory was eagerly seized upon by politicians, social workers, economists, etc., to interpret and predict events in various spheres. As Irvine notes:

"Kant and Laplace found it in the solar system, Lyell on the surface of the earth, Herder in history, Newman in church doctrine, Hegel in the Divine Mind, and Spencer in nearly everything." [16 p67]

Today, evolution dominates much of the philosophy behind our Western thinking. Even in religion, Jewish monotheism is said to have "evolved" from paganism and polytheism – an assertion which is blatantly false, for the Bible is clearly monotheistic from beginning to end.

THE RIGHT WING

Regarding the politics of the Right Wing, it is a well known fact that Hitler based his theories of race and war upon the writings of Nietzsche, who had declared that wars are justified, as the strongest race would conquer the weakest. Conversely, Christianity was to be destroyed as being counter-evolutionary. With such a basis, the use of the gas chambers in the Concentration Camps is a logical outcome.

THE LEFT WING

Similarly with the Left Wing, the immediate indoctrination of conquered territories by enforced lectures on evolution as well as Marxist politics is well recorded. The outcome of the brutalizing effect which such teachings produce is frightening in its revelation of how soul destroying such a process can be.

One of the most important books which documents just how terrible are the consequences when a political system is unrestrained by any motives of justice or kindness is entitled *The Dark Side of the Moon*[33]. The book, which is based upon first hand accounts of communist brutality in the deportation and treatment of Polish nationals, is an appalling document of the inhumanity to which godless men can descend. The authoress (who is not named) had access to the papers of the Polish General Sikorski and the book carries a foreword by T.S. Eliot.

It is quite sickening to read of the mass transportation in cattle trucks, the gross overcrowding of the prisons and the callous starvation and annihalation of innocent ordinary Polish people and members of the armed forces. The total effect is numbing and caused the author to seek for an answer to the question of how men could become so completely devoid of even a spark of humanity. She gives an account of the attitude of the cold and callous guards on one of the trains deporting people to the Russian interior and continues:

"I have searched in vain through masses of evidence for records of anything approaching humanity being shown by any soldiers on any of these trains. I have a record of one man passing in an extra bucket of water and five other records of doors being opened for a short period, some ten minutes or so, to let some air in; and this after the most urgent entreaty and from caprice, not in any way furnishing a precedent for other occasions. This question has preoccupied me profoundly. Throughout my work on this book, work which has occupied several years, I have searched the evidence exhaustively on this point. I have also put the question to every single person with whom I have talked. It has been of immense importance to me, of an importance greater than I can possibly express, to discover that some instinct of humanity did survive, somehow. The answer invariably given me has been that it did not" [33p67].

We are frequently reminded of the horrors of the Nazi concentration camps but it would appear to me that the conditions related in this book are even worse. Yet so little publicity is given to these planned and systematic atrocities of the communist regime that few people know that they even took place.

Here indeed is the brutalizing of the human mind which Sedgwick was quick to perceive. When it was later pointed out to Darwin that he was providing every criminal with a means of justifying his actions, he merely dismissed the idea as a "good squib". He clearly had no wish to be drawn into a discussion regarding the appalling consequences which could result from the application of his theory in the sphere of morals and ethics.

Destabilisation

There is one addional aspect of evolution which strikes at the very root of all societies where a basic stability is an essential factor. The continuing propaganda of evolution makes it far easier to proclaim that "changing patterns of life" are all part of the evolution of society. As "evolution is progress" then "change" is presented as "progress". Thus our culture is now subjected to a bewildering series of changes, often for no apparent gain. Our society becomes increasingly destabilised and thereby more open to those who proclaim that only a strong "democratic" centralised rule can provide the citizens with the peaceful life which the vast majority now long for. Thus is the way prepared for a change in our political system, whether by violent or 'peaceful' means.

THE GODLESS "FAITH"

The important question is – why should the theory of evolution, which purports to be a scientific explanation of how species developed, have such drastic repercussions in the field of human relationships?

The answer is surely quite simple. Deeply planted within all men is the idea that there are ultimate standards of right and wrong which have been set by a God who is infinitely superior to man himself. As C.S. Lewis has shown in the first chapter of his famous book, *Mere Christianity* [30], it is to this ultimate standard that men automatically appeal when justifying their actions. Thus to say that a certain action "is not fair" means you are claiming it is unjust, and thus falls short of a standard of perfect justice to which all men appeal when arguing. To ignore completely any regard for justice and fairness in behaviour is to degrade all human civilization to a savage chaos in which self-survival would be the most important characteristic.

What is important is that this standard is only accepted if there is a tacit acknowledgement that it is outside of and superior to man himself, having been imposed by a being who is himself outside of and superior to mankind, i.e. God. Standards of right and wrong deal with relationships between people and it would be ridiculous to claim that they exist 'on their own'. They are clearly imposed by a personality who is superior to man. The essential point is that it is the concept of a God who is able to set such standards of perfection, which gives meaning to the standards themselves.

The disastrous effect of evolution is that it allows a philosophy to flourish which gives a purely materialistic explanation of how life arose. It thus dispenses with the need of a God. *Once God has been eliminated, then any basis of absolute moral standards is destroyed, and all behaviour becomes 'relative'.*

OUTWORKINGS

This dismissal of God completely removes the ultimate authority which could regulate our social conduct. To illustrate just how completely undermined is the basis of our moral actions, the following examples are provided.

i) Conscience

In order to persuade people to act fairly and justly one could appeal to their 'conscience'. But in a purely materialistic-evolutionary world there is no such 'thing' as 'conscience', and it cannot therefore be appealed to.

ii) National pride

In times of war, people may be exhorted to be prepared to die for "the good of the nation" on the basis that the sacrifice of a few individuals will permit the survival, dominance and future progress of the nation. But such an argument *has no more validity* than that of an individual who says: "I am following the l aw of survival of the fittest when I evade call-up, for by doing so, I am less likely to be killed. I will then be able to have offspring who will be as 'clever' as I am at surviving"!

iii) Criminals

Theft can easily be justified by such excuses as "spreading wealth more uniformly" or even perhaps by the fact that, as there are parasites in the animal and vegetable kingdoms, why should we object to their existing within the human species?

Continuing this further, there are no ultimate objections to one man killing another on the slightest pretext. Thus we see wars and the mass slaughter of whole races being carried out without the least twinge of conscience on the basis that "might is right". When the fear of a God who will one day judge our actions is absent, there is no limit to the means by which man will fulfil his self-centred ambitions.

iv) State psychiatry

I have already referred to the social pressures which make us conform to certain attitudes and how non-conformity can result in isolation. This can be taken further, as intimated by Garret Hardin of the Californian Institute of Technology, who said that the man who does not honour Darwin "inevitably attracts the speculative psychiatric eye to himself"[34]. In this way the seed is sown of the idea that those who are not evolutionists are mentally abnormal and therefore are in need of treatment. This principle can be extended to any who do not conform to what the "state" considers "normal".

Such a seed has indeed germinated and flowered in all its horror in the psychiatric wards of the USSR, to which Christians, conscientious objectors, human rights campaigners, and many other such groups, who act contrary to the state's wishes, are sent for 'treatment'.

As I write, I have before me an account of a visit to a young Christian by his fiancee and a friend who spoke to the psychiatrist, Vladimir Levitsky, about his case. The psychiatrist, who boasted that he was "an atheist through and through" made such comments as:

"...you are talking about him as if he were a healthy person, and I find that he is ill....Religion for him is an idee fixe....As an atheist, I consider it [faith] an abnormality...even an illness....he has to be treated and we are going to do it....we shall treat him with medicine....I shall keep him in hospital as long as necessary....We are tearing Eduard's personality apart! You are pulling him towards God, and we...towards the Devil...so I am using my rights as a psychiatrist to deny you and your friends access to him. And I personally request you to leave him altogether".

In reading such accounts, one cannot but have the greatest fear for the sanity of the broken wrecks who survive such treatment...if they ever do.

In giving these four examples of how "survival of the fittest" can be used to justify the most anti-social of actions, it may be objected that, whilst the vast majority accept evolution as a fact, this does not result in their acting in the completely unmoral ways I have outlined above. Indeed, most profess the very highest of motives in their conduct.

The correctness of this statement is not disputed. The whole point, however, is that this sense of fair play, conscience, call it what you will, does not, indeed cannot, come from the evolutionary theory to which they hold. Thus for most people, the theory of evolution is only a mental assent, but what controls the morality of their actions is still that small voice of conscience which is deeply implanted within them by God.

It is possible to highlight this dichotomy by discussing the subject with evolutionists, who will act with the greatest charm and the highest of motives in their dealing with their friends. If challenged on why they should act thus in view of their philosophy, they may recourse to such excuses as "beneficial to the race" or "social convention", but as C.S. Lewis has shown [30], all such reasons can be traced back to a concept of some absolute standard of right and wrong.

This concept of "fairness" or spark of conscience is very deeply planted within the hearts of men. In some brutalized individuals it seems to be either ignored or even non existent. Each man is ultimately responsible for the extent to which he listens to or ignores this inner voice, irrespective of any detrimental factors in his environment or upbringing, etc. Having said this, however, it is obvious that the state, mainly by its attitude to religious principles, has enormous power either to encourage or discourage men to act in accord with their conscience. Indeed, the subject of the link between a nation and its religion is of primary importance.

CHAPTER 34

NATIONS AND THEIR GOD

That very incisive Christian writer, A.W. Tozer, once wrote:
"The history of mankind will probably show that no people
has ever risen above its religion, and man's spiritual history will
positively demonstrate that no religion has ever been greater than
its idea of God. Worship is pure or base as the worshipper
entertains high or low thoughts of God.... We tend by a secret law
of the soul to move towards our mental image of God.... Were we
able to extract from any man a complete answer to the question,
'what comes into your mind when you think about God?' we
might predict with cetainty the spiritual future of that man. Were
we able to know exactly what our most influential religious
leaders think of God today, we might be able with some precision
to fortell where the Church will stand tomorrow"[35p9].

From this, it is a simple step to realise that a nation's concept of
the nature of God will ultimately shape the pattern of culture and
civilization which that nation pursues.

It does not take much imagination to make a broad survey of the
concepts of God which the various nations of the world have, to
realize that there is a close correlation between this concept and
what we will call the "quality of life" which that nation enjoys. To
emphasize this, I would ask, who, other things being equal, would
prefer to be an ordinary citizen of a Far Eastern country rather than
one of an industrialised European nation?

Quality of life
It may be asked, what do I mean by "quality of life"? I would
broadly define this as freedom of the individual to direct his life with
a minimum of restrictions and in an atmosphere of trust and
goodwill towards his fellow men. Signs of such conditions existing
would be:

a) a low crime rate
b) considerable charitable giving
c) provision of national institutions (hospitals, old people's
   homes, etc.)
d) easy access to legal justice
e) absence of corruption at all levels
f) low incidence of divorce, immorality, etc.

— and many others could be added. Indeed the reader may like to extend the list.

It is also obvious that those nations with many of the features listed also maintain a high standard of living. The reason for this is surely that, in a nation as in a business community, when there is mutual trust and respect, this freedom from internal "infighting" to maintain one's position in life allows all the available energy to be channelled into creating greater wealth in which all can participate. This rise in the welfare of certain nations is so marked that social historians have given it a specific label – the "Protestant work ethic" – clearly an acknowledgement from an unexpected quarter of the link between a nation's faith and its welfare.

The root of the present day high level civilization sprang from those nations where there was a sound Christian basis. Indeed, it is surely evident that there is a very close relationship between the 'quality of life' enjoyed by a country's citizens and the percentage of them who hold to a scriptural faith. Even geographically, one could trace a curve running through Switzerland, Germany, Holland, England and ultimately America, of countries who have a reasonable percentage of 'born again' Christians. What is significant to me is that the founding fathers of these nations held to the basic doctrines of the Christian faith shared with Augustine, Calvin, Luther, Kuyper, the Puritans, Wesley, Wycliffe and Whitfield.

Theological roots

What is the root cause for this link between a nation's life and its God? Surely, when a man (and it comes down to individuals in the end) acknowledges that he will one day have to answer to a God who is eternally righteous, just and yet loving, he will willingly obey His commandment that he should deal likewise with his fellow man, for he is similarly made "in the image of God". His religious faith will affect his attitude to life and, where there are many similarly motivated individuals, the whole of society will be permeated by such views, resulting in a 'Christian Nation'. Doubtless, many will be 'nominal' Christians, paying lip service to religious observances and following the moral standards of their social class. But such standards are set by those for whom their religious faith is the dominant factor in their lives. In this situation, 'nominal christians' are merely trading on the spiritual inheritance which their forbears paid for with their lives.

Other religions also have high codes of behaviour which they exhort their believers to follow. Yet their failure to affect the lives of individuals in the same way that Christianity does is obvious. It is here that the marked contrast between Christianity and all other faiths is displayed, for it is the only one which claims that God has taken the initiative by coming to earth and directing men's lives. It is His power, and His alone, which can possibly change men's hearts, and ultimately their lives, both temporal and eternal.

The effect of evolution

Having read so far, one may well be asked, "What has all this religious survey got to do with the 'scientific' theory of evolution?"

The purpose, at a risk of over emphasis, is to make it clear how fundamentally important to a nation is its concept of God. *Anything which detracts from this in any degree, will ultimately damage the fabric of a nation's life by a corresponding amount.*

No nation changes its religious attitude or philosophy rapidly, except by revolution. It is invariably a slow process of many generations of rising, despite bitter persecution, and declining due to apathy. The decline, however, is rapidly accelerated when the foundations are undermined.

With this in mind, just how damaging the long term effect of the teaching of evolution is will now perhaps be appreciated. To present a nation with a false philosophy which enables it not just simply to demote God, but to banish Him entirely, is a major achievement of anti-Christian forces.

Perhaps it is only now that the importance of Lyell's long term aim of destroying the "Mosaic account" by "forty years' march of honest feeling" will be fully appreciated. The results are now visible not only in this country but throughout the world. This is not to say that were evolution eradicated, all would be well with the world — man's antipathy to God runs far deeper and he would develop some alternative system to justify his Godless attitude. I do contend however that in Western society at least, it is one of the main philosophies which has undermined the Christian foundation of many nations.

CHAPTER 35

CREATION v. EVOLUTION - THE BALANCE

The question is sometimes asked "If God did create the world as believing Christians assert, why then did He leave so much 'evidence' which seems to support the theory of evolution?"

It is also noted that when some of the available evidence is examined, *it is possible to interpret it in either a Creationish or an Evolutionist way!*

The following examples show how this dual interpretation operates.

1) FOSSILS

Evidence

In the majority of geological strata are found the fossilized bones of a vast number of animals, sometimes in small pieces, at other times the whole specimen is virtually intact.

Evolutionist interpretation-

Over millions of years, some of the animals (or plants) would fall into rivers or lakes and be eventually covered with sediments. This would preserve them from decay and fossilization would take place slowly over many years.

Creationist interpretation-

All living creatures were swept to their death during one cataclysmic flood. Burial was rapid, and fossilization would not have taken millions of years.

2) STRATA

Evidence

When the geological strata are studied, there is *in general* an appearance of 'development' of the animals whose fossilized bones they contain, as you go up through younger strata (See Appendix Fig. 1). Thus in the lower strata (Cambrian) you have only [?] Invertebrates. Further up the column, the first fishes appear, then the first amphibia, later the reptiles, then mammals and birds, with man first appearing only in the highest strata.

Evolutionist-

As life evolved, the bodies of some of the species then living were buried, their fossilized bones being effectively a random sample of the species existing at the time when the stratum was laid down. There are bound to be large gaps in the series of fossils as the geological strata are very incomplete, most strata suffering from further erosion and redeposition.

Creationist-
    What appears to be an 'order of development' is really the order in which creatures would be inundated by a catastrophic flood. First the slow moving sea creatures would be covered by the flood sediments. Fishes, which could swim faster, would nevertheless eventually be overwhelmed and suffocated by the muddy waters. Amphibia would be the first land creatures caught by the rising waters and later the slow moving reptiles. Birds would eventually fall exhausted into the water, and the larger, faster land mammals would be amongst the last to succumb. Man would keep afloat the longest, exercising all his ingenuity in obtaining support from any available floating material. Thus the 'order' of animals in the strata is a reflection of their ability to survive a world wide flood, rather than the stages of animal evolution.

3) HOMOLOGIES (Similarities between species)

Evidence.
    An examination of living species shows that, although many of them appear to be quite different, they may nevertheless have much that in common. This is particularly so regarding their skeletal frame, for a vast number of species have a backbone, a rib cage, a skull and four limbs, each with five digits at the end.

Evolutionist-
    As life evolved, the basic skeletal framework appeared and this was capable of being adapted and modified advantageously under a wide variety of conditions. The fact that so many different animals have the same skeletal pattern is clear evidence that they have all evolved from a common ancestor.

Creationist-
    When God created the various animal forms, He used a basic pattern which was perfectly adequate for a wide range of variations between similar shaped species. There was no need to make a completely different framework for every species. God simply exercised economy and efficiency in his various designs. Furthermore, whilst very similar looking animals may appear to be closely related, detailed investigation of various skeletal and biological

features shows that, nevertheless, they are really very dissimilar, and *not* closely related.

These three examples should be sufficient to illustrate my point. Similar examples could be given from theology such as the inability to either prove or disprove the existence of God and that we have both love and hatred in the world.

These examples show how evenly balanced the two possible interpretations would appear to be to an unbiased observer. We must now return to the fundamental question: If the Creationist view is correct, why should God have permitted so much evidence to exist which supports the theory of evolution - a theory which dismisses Him as an active agent in the Universe?

To answer this we will examine two alternative situations which God might have allowed, then show the theological self contradictions which then arise, and finally give an explanation of why He deliberately allowed the present situation to exist.

1) ALL CREATION

If God had so willed, He could have designed the world so that there was no spark of evidence which could possibly support evolution. In such a situation, by using their normal reason, men would have no option but to acknowledge that an infinitely clever creator-God existed. He would thus be effectively imposing the evidence of His vast power upon the senses of all men, from which they would have no escape. From this it might be said that God was effectively 'forcing' men to worship Him as the only true God.

2) ALL EVOLUTION

Alternatively, God could have left no evidence of Creation, but much which supported evolution. In this case again, any man, using his normal powers of reason, would conclude that evolution had taken place. In this aspect (of science at least) he would have no cause to believe that God exists. (For the purposes of this argument, we will ignore the other means by which God may reveal Himself to a man, such as personal (spiritual) knowledge, miracles, the Bible, etc., and confine myself to the purely scientific evidence in order to draw a clear line for clarification of the issues involved.)

If we assume there was no evidence which suggested that a powerful creator-God existed, then any man who nevertheless wished to believe this to be true, would have to flatly contradict his normal reasoning powers as he considered the evidence before him. But in a perfect God, perfect reason is one of His many attributes. He could not therefore logically "create man in His image" and then expect him to act in complete opposition to the gift He had given him, i.e. having given him Reason, He would not then expect him to act unreasonably or irrationally in order to believe God existed. Such an action by God is inherently self-contradictory, and God is certainly not that!

These two possibilities have been deliberately exaggerated to emphasize the fundamental issues they would raise. Why then did God allow the present seeming dilemma to arise?

THE BALANCE

We have then an apparent balance between the evidence supporting creation and that for evolution. As we have seen, we cannot carry out an 'experiment' of origins to determine which is correct. We are limited to interpreting the evidence before us. Any interpretation depends upon our rational thinking process, but as I have shown, two equally valid interpretations (superficially at least) are possible.

From this it is obvious that it is not finally our reason which really decides what we believe. If this is so, what then does determine our belief, and why has God left two feasible inter-pretations? I would suggest the answer to these two questions is as follows:

God has deliberately not forced us, when we use our reason, into either of the two views, and by inference into the particular view of him which they imply. Our view of God (our religion) is not a matter of reason, but above all, a matter of the heart.

Thus we are faced with a choice, but the result depends not upon a rational examination of evidence, but upon our religious view of life. Those who are atheist would claim that they do not have any religious views but only a philosophy of life. However, the tenacity with which they hold to them and the emotion that they can give rise to suggests that their roots are far deeper, for they are the basis of a 'faith' which has excluded God. As a friend commented, "Be careful when you are criticising someone's philosophy of life. You are attacking his religion and you will have a fight on your hands!"

THE CHOICE

In the previous section, we looked at the important effect which Conceptual Frameworks have upon those who hold them in their interpretation of scientific evidence. In this section we are consid-ering not just a specific Conceptual Framework regarding a a scientific theory, but what might be called the Total Conceptual Framework, which every person has in his interpretation of the Meaning of Life.

This Total Conceptual Framework is ultimately determined by

our religious view of life. To put it in its correct order, our view of God is the most important determining factor in the way we look at life (or erect our Total Conceptual Framework), as we have quoted A.E. Tozer above. It is then this Total Conceptual Framework which determines our views and reactions to the normal everyday events around us.

There are three main forms which these frameworks may take, which are fundamentally determined by our view of God.

A. The Godless view

There is nothing except matter, nothing outside nature and no ultimate moral law. As we have seen, with this view there are inherent logical contradictions and that furthermore, those who profess this view will borrow the concept of perfect justice from the Religious view when they appeal for fairness in other people.

B. The religious view

The existence of God is acknowledged but views of His precise relationship with the workings of the Universe and in particular with man will vary from individual to individual. This group would on one hand range from the Deist (who says that God is letting the world run its course without interference) to the Theistic Evolutionist, who would say that God put a soul into a highly evolved form of ape, thus making him Man. I would suggest that this group of views is the result of an inadequate concept of God, each individual giving as much prominence to God's activity and ability as he so wishes. Not unlike the Life-Force view, God can easily be ignored when the individual wishes to undertake some 'un-Godly' action.

C. The Sovereign God

God is infinite in power and wisdom. He created the whole Universe and placed man in it as its head, giving him a record of his creative acts in His Word. The Universe operates in accordance with the Laws which God has laid down, and which man may explore and discover. The existence of evil in this world is the result of man's subsequent rebellion against God.

THE ROOT OF OUR VIEWS.

These are the three basic forms which a man's total Conceptual Framework may take, each one depending upon his religious view of life and not initially upon his rational thinking and logical deduction from observed facts. To state this in another way:

We all determine first what is our world view of life and then will only accept as 'facts' that evidence which is in accordance with it.

CHAPTER 36

INADEQUACY OF FACTUAL EVIDENCE

If the case set out in the previous chapter is accepted, then one important issue stems from it. This is that it is impossible to convince an evolutionist, whether Theistic or Atheistic, of the falsity of his views on purely factual evidence alone! It will be obvious that before any man can undergo such a fundamental change, involving as we have seen his total view of life, he must *first* greatly enlarge his concept of God. Thus, arguments between the two viewpoints invariably hinge on what is acceptable as adequate 'evidence', for the basis on which they assess this is completly different. It is therefore obvious that before anyone can accept the Creationist view, *there must first be a change in that persons Conceptual Framework*. The new framework must at the very least include a God who is powerful enough to create whatsoever He wishes in an instant of time. It is this limited conceptual framework which prevents many from accepting the creationist viewpoint.

Purpose of creationist efforts

If an evolutionists view of life is not changed simply by the presentation of evidence, this does raise the question "Why do creationists lecture and write books dealing with the factual evidence, if ultimately such facts cannot effect a change?" The answer to this is threefold.

1) ENCOURAGEMENT.

It greatly encourages those Christians who have already acknowledged that the Bible is spiritually true, to be shown that it is scientifically true also. It gives them even greater confidence that the Word of God is totally reliable.

[The work of archaeologists in proving its historical accuracy - despite the claims of error by liberal theologians - is performing a similar service in this area also. An important book on this topic is "The Exodus Problem and its Ramifications" by D. A. Courville [56]. In this the author shows (as does Velikovsky [57]) that the accepted datings of Egyptian chronology are far too long. When they contradict the Biblical accounts of history, inevitably it is claimed that the Bible is wrong. However, Courville presents a well documented case for the Egyptian time scale to be drastically shortened, and when this is carried out, the agreement with the Biblical account is remarkable].

2) COUNTER-EVOLUTION

Not infrequently, a young Christian will accept the Bible as true, but when he attends school or college, he is presented with the 'facts' supporting evolution and with no evidence which supports a Creationist interpretation. This results in considerable confusion and he naturally begins to doubt the Genesis account. This can then lead to doubt about the accuracy of the Bible in other matters and can occasion complete loss of faith. It is to prevent such an unwarranted deterioration of faith that the work of presenting the Creationist case should be carried out. Those who lecture on the subject can testify to the gratitude expressed by many Christians when they hear for the first time the evidence supporting the Bible which they have taken on trust for many years despite the immense evolutionary propaganda to which they have been subjected.

3) PUBLICITY

The general public, whether church-going or not, need to be informed that there is a perfectly rational creationist interpretation of the Universe, which is scientifically sound and 'respectable'. As I have said, facts cannot change concepts *on their own*, but can prepare the way for the necessary spiritual change. It is the duty of all Creationists to plough, sow and water, in the hope that in some cases at least God will bring about a rich harvest.

TESTIMONY

Before closing this section, there are two further points to which I would like to refer, both being in the nature of a testimony.

1) THE FAILURE

When I first investigated the evidence against evolution, it became clear to me that the theory was little short of a vast scientific hoax. I studied the subject in some detail, as I felt sure that it provided a method by which I could 'prove' by scientific means that God existed. I envisaged the situation rather like the diagram in figure 14

In this, each person is faced with the choice of Evolution or Creation. Those who accepted creation would obviously acknowledge that a Creator-God existed. Evolutionists have a further choice of either denying that God existed, or if He did exist, that He controlled the course of evolution i.e. Theistic Evolution. I fondly imagined that if I presented sufficient scientific facts to sensible rational people, they would have to admit that evolution was wrong, and that conversely, the evidence for creation 'proved' that a Creator existed. Thus, by cutting the options on the diagram at X, I thought that I could persuade people to acknowledge the existence

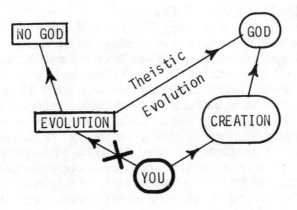

Fig.14. 'The Choice' diagram

of a Creator God.

It was after many months of lectures, discussions with graduates, etc., that I came to realize how ineffective was the pure presentation of facts. It was then that the principle occurred to me which I have given above, i.e. that on a subject as fundamental as evolution (vs. Creation), people make their minds up first and then will only accept the facts which comply with their views. The lesson I learnt was both important and chastening.

2) ASSURANCE

Those who hold to the creationist view are often labelled as "fundamentalist bigots who cling to a narrowminded interpretation of the Bible, despite the scientific proof against it."

I would insist that my realization (and I am sure that of many of my Christian readers), that the Genesis account of Creation is accurate, results in an immense *broadening* of one's vision regarding God and His plan. Simply to conceive in our very limited way, just a small part of the immensity and power of God, is absolutely breathtaking in its awesome splendour, and puts one aspect of our relationship to such a God in its true perspective. This heightens even further the realization that this same all-powerful God should nevertheless be so infinitely humble that He was prepared to take the punishment of death in place of His rebellious creation — man —, in order to restore him to the original relationship of love between them.

When such a concept grips the mind, the effect is overwhelming. When furthermore it is supported both by an inward spiritual witness and by external evidences, there comes a very strong sense that one has discovered the source of all Real Truth at long last. This is not to claim that our imperfect minds do not at times err, but they can always return to the solid foundations of God's truth.

It is this unshakable inward conviction which so greatly offends the opposition and which they are quick to label as "bigotted". However, when one is utterly convinced that you have found the Truth, any movement can only be towards error, in the same way that a man at the North Pole can only travel South.

When accused by well meaning people of being "narrow-minded", or told that we should "keep an open mind", I have a picture of someone swimming in rough water amidst uncharted rocks, insisting that those of us who are on firm ground should join them in their "flexible positions".

In answer to such calls, we, who have found this rock of Truth, can only reply, as did Luther when similarly charged:

"Here I stand. I can do no other".

CHAPTER 37

THE FUTURE

In England, as in most countries, there has been a visible decline in the "quality of life". The very considerable increase, over many years, of crime, industrial unrest, divorce, etc., is well documented and beyond dispute. The Christian faith is gradually being demoted, particularly in schools, where it is taught in Comparative Religion as just one of many world faiths along with Buddhists, Muslim and Hindu religions. Under the heading of "Combined Studies", evolution in its broadest interpretation is the basic concept which is used to link the various subjects of Biology, History, Sociology, Economics, etc., and to show how they have each "developed".

Studies have shown that all the major civilizations have really declined or fallen due initially to internal corruption and decay of the numerous threads which hold a complex organization together. Once the will of a population to abide by sensible laws has declined to that of selfish interest which they see exhibited in the leaders of their country, then their days are surely numbered. Gibbon's *Decline and Fall of the Roman Empire* is a catalogue of the fate of just one such history. In the end days of that nation, the people were kept distracted from the growing problems by the provision of "bread and circuses". One cannot but wonder, as we watch the media in action, whether their counterparts today are Sex and Football!

In the international sphere, there are "wars and rumours of wars". Terrorism increases, and no man feels safe.

Surveying the whole scene, many Christians have a strong sense that God is slowly removing His restraining hand, allowing man more freedom from the limits to his self aggrandisement which God has generally imposed upon mankind for its own good. As God relaxes His control, man is left free to yield to the self-centred desires deep within the heart of every one of us. Christians who can see this rise of malignant forces throughout the nations can only stand horror struck as they watch the world slide into chaos exactly as predicted in the last book of the Bible.

Beneath our feet lie the millions of fossils of extinct animals and men. It is a silent witness to a time when mankind became so degenerate that finally it was virtually wiped out in one great cataclysmic flood. Fearful though this thought may be, we can take some comfort from the fact that we have been promised that this will never happen again.

Next time it will be by fire!

CHAPTER 38

CONCLUSION

Those readers who have reached this point may well feel a little bemused from following the wide range of subjects touched upon in tracing the ramifications of the theory of evolution. We have examined its early history, its rise in the mid nineteenth Century, its use by both political extremes, its philosophical weaknesses and its theological implications.

In particular, I have presented the evidence which suggests that far from being a scientific discovery, it was a deliberate imposition of a false scientific dogma upon the unwitting public of the day by a comparatively small group of people. Whether or not there were other forces behind them I leave as an open question. I am not of course suggesting that everyone who actively promotes evolution is involved in some form of conspiracy for the majority are unaware of its implications. I do suggest that anyone who believes the theory, particularly if they are involved in teaching it, should recognise its scientific inadequacy and the disasterous effect of its outworkings in the life of the nation.

What I hope is reasonably substantiated by this book is that above all else, the theory of evolution is fundamentally an attack upon Christianity. This can be traced from Lyell's deliberate undermining of the "Mosaic account of Genesis" through to the enforced indoctrination of the subject in Marxist dominated countries, which are deliberately trying to eradicate the concept of God. This being so, I am quite certain of the ultimate source from whence come all such false concepts and deceptive theories. This source was neatly described in a comment made by a friend of mine whilst we were discussing the subject of evolution. I had been describing the very dubious circumstances surrounding the fabric-ation of the ape-men 'missing links'. He paused for a moment, and then quietly said:

" Hmmm....

...you can smell the sulphur!"

APPENDICES

APPENDIX I

A RETURN TO FAITH ?

The possibility that Darwin returned to the Christian faith in his last years has interested many people although denied by his supporters. The main source of this information is given in pamphlet No. 80 of the Evolution Protest Movement (now Creation Science Movement). Although other accounts have appeared in some American publications they are very similar to that in the pamphlet, which, as it seems to give the fullest account will be used as a basis for this appendix.

A Lady Hope is said to have visited Darwin shortly before he died and gave an account of their meeting at the evangelist, D.L. Moody's educational establishment at Northfield, Boston. Her written account appeared in the *Boston Watchman Examiner* and was later reported in the *Bombay Guardian* of 25 March 1916.

Lady Hope's account

In her account, Lady Hope said:

'He waved his hand toward the window as he pointed out the scene beyond, while in the other he held an open Bible, which he was always studying.'

"What are you reading now?" I asked, as I seated myself at his bedside.

"Hebrews!" he answered, "still Hebrews, the Royal Book, I call it. Isn't it grand?"

'Then placing his finger on certain passages, he commented on them.'

'I made some allusions to the strong opinions expressed by many persons on the history of the Creation, its grandeur, and then their treatment of the early chapters of the Book of Genesis.'

'He seemed greatly distressed, his fingers twitched nervously, and a look of agony came over his face as he said:'

"I was a young man with unformed ideas. I threw out queries, suggestions, wondering all the time over everything and to my astonishment the ideas took like wildfire. People made a religion of them."

'Then he paused, and after a few more sentences on the holiness of God and the grandeur of the Book, looking at the Bible which he was holding tenderly all the time, he suddenly said:'

"I have a summer-house in the garden, which holds about thirty people. It is over there," pointing through the window. "I want you very much to speak there. I know you read the Bible in the villages. Tomorrow afternoon I should like the servants on the place, some tenants and a few of the neighbours, to gather there. Will you speak to them?"

'"What shall I speak about?" I asked.'

"Christ Jesus," he replied in a clear emphatic voice, adding in a lower tone, "and His salvation. Is not that the best theme? And then I want you to sing some hymns with them. You lead on your small instrument, do you not?"

'The wonderful look of brightness and animation on his face as he said this I shall never forget, for he added:'

"If you take the meeting at three o'clock this window will be open, and you will know that I am joining in with the singing."

'How I wished that I could have made a picture of the fine old man and his beautiful surroundings on that memorable day.'

This account has been flatly denied by evolutionists, and even some creationists have doubted its authenticity, notably W.H. Rusch Snr., a Dean of Concordia Lutheran College, Michigan. He points out that there is no indication of "recantation" or resurgence of Christian faith in any of Darwin's letters, either in the years before his death or even in those written a few months before he died. Rusch also checked the available numbers of *The Watchman Examiner* and could find no mention of the account. Similarly he could find no trace of Lady Hope. He concludes that the whole account is fictitious and therefore liable to do more harm than good to the creationist cause. He does not name any possible supects but wonders whether Darwin's wife, Emma, was behind the story. She had tried to censor some of his letters when published after his death, for she did not want Darwin to appear to say that man's spiritual views were no "higher" than their animal origins. Rusch admits, however, that as a Unitarian she would be unlikely to be connected with an account in which Darwin returned to the Christian faith and therefore became a Trinitarian.

Clearly it would be difficult to prove whether the meeting took place as recorded. There is, however, some very interesting additional evidence regarding Lady Hope, whose identity, until recently, was as nebulous as the authenticity of the account itself.

Lady Hope's identity

In April 1979, Dr. C.E.A. Turner, Chairman of the Creation Science Movement, received a letter from Mr. L.G. Pine, B.A., who was at one time the editor of *Burke's Peerage*. In this letter he said:

"Now with regard to Lady Hope, I think that I have uncovered her identity, which should be a help to tracing the story.... Under the article in B.P. for Viscount Combermere I found a mention of Elizabeth Reid Stapleton-Cotton, whose date of birth is not given (convention of those times), but who was born soon after 1841. She married in 1877 Adm. of the Fleet Sir James Hope, G.C.B. She was obviously much younger than he. He d. in 1881, and she married again, in 1893 a Mr. T.A. Denny. She preferred to be known as Lady Hope right up to her death in 1922.

I think this is the Lady Hope in connection with Darwin, as no other at that time, i.e. round 1882 will fit."

The fact that a Lady Hope has been identified who was in the same social class as the Darwin family certainly makes it more likely that the encounter did take place even though no direct connection has yet been established. There is however other evidence that in his closing years Darwin did experience a renewal of his Christian faith.

Further evidence

Darwin's acknowledgement of the effectiveness of evangelistic preaching in changing men's lives is provided by the following incidents. These are given in two books which record the activities of members of the Brethren Movement entitled *Chief Men among Brethren* [48] and *A History of the Brethren Movement* [49].

a) FUEGANS

During his voyage on the *Beagle*, Darwin formed a very low opinion of the inhabitants of the island of Terra del Fuego, considering them too low and primitive in their ways for any attempts to raise them to succeed. His acknowledgement that he was wrong is recorded as follows:

"Admiral Sir James Sullivan wrote to the *Daily News* on April 24th, 1885: 'I have been closely connected with the South American Missionary Society and Mr. Darwin had often expressed to me his convictions that it was utterly useless to send missionaries to such a set of savages as the Fuegians, probably the lowest of the human race.'"

"'I always replied that I did not believe any human being existed too low to comprehend the simple message of the gospel of Christ."

"'Many years after he wrote me, now saying that recent accounts of the mission proved to him that he was wrong and I was right and he requested me to forward the enclosed cheque of £5 as a testimony of the interest he took in the good work.

"On June 6th, 1874, Mr. Darwin wrote: 'The progress of the Fuegians is wonderful, and had it not occurred, would have been to me quite incredible.'"

"In March, 1881, he wrote: 'The account of the Fuegians interested not only me but all my family. It is truly wonderful what you have heard from Mr. Bridges about their honesty and their language. I certainly should have predicted that not all the missionaries in the world could have done what he has done.'"

"It will be remembered (from our 1958 Centenary Supplement) that Darwin had been utterly misled by his evolutionary ideas as to the mentality, morality and language limitations of the Fuegians. The fact is, primitive man is an evolutionary myth. Man is not a rising beast, but is rather a fallen son of God, who cannot raise himself. Yet with God's help he is capable of the great improvement he now so obviously needs, capable of being, with Disraeli, on the side of the angels" [48]."

b) DOWNE MISSIONS

The Darwin family was particularly impressed with the results of the evangelistic work of a Mr. J. Fegan who was active in Downe village.

"Mr. Darwin rented an old schoolroom in the village of Down for the use of the villagers. Mr. J. W. Fegan asked for the use of the hall to preach the gospel and was always given permission. On these occasions the Darwin family were considerate enough to alter the dinner hour so that any who wished to, might attend. Passlow, the family butler and Mrs. Sales the housekeeper both were converted. Mr. Fegan wanted the hall for a week's mission so he wrote Mr. Darwin for permission and his reply was as follows:

'Dear Mr. Fegan,

You ought not to have to write for permission to use the reading room. You have far more right to it than we have, for your services have done more for the village in a few months than all our efforts for many years. We have never been able to reclaim a drunkard, but through your services I do not know that there is a drunkard left in the village. Now may I have the pleasure of handing the Reading Room over to you. Perhaps if we should want it some night for a special purpose, you will be good enough to let us use it.

Yours sincerely,

CHARLES DARWIN.'"

"The transfer was duly made, the Reading Room was then called "The Gospel Room" and services have been held there continuously for half a century" [48]

A separate account has been written of Fegan's work and the approval which this received from the Darwin family:

"Fegan's work gained the sympathetic interest of the Darwin family at Down in Kent, to which village his parents had moved their home, and it was through Darwin's goodwill that Fegan was able to obtain the use of the village Reading Room for services. Members of the Darwin household attended his services, in which the family itself maintained a sympathetic if distant interest. They were apparently particularly struck by Fegan's success in reforming alcoholics. Not only did Charles Darwin himself comment on this, but Mrs. Darwin wrote in the course of her family correspondence:

'Hurrah for Mr. Fegan! Mrs. Evans attended a prayer meeting in which old M. made 'as nice a prayer as ever you heard in your life' [49p180].' ['Old M' was a notable drunkard in the village of Down, converted by Mr. Fegan.]"

All the evidence presented above certainly shows that in his personal dealings at a local level he was at least very sympathetic to Christianity and far from being antagonistic. What then do we make of the evidence that he did *not* return to Christianity? Thus Rusch points out that there is no hint in Darwin's published letters that he had had second thoughts about Christianity and that the account may have been fabricated.

Similarly we find that in his letter to Dr. Aveling in 1880 and at his meeting with him in 1881, Darwin expressed strong anti-Christian views as we have discussed in chapter 23. One cannot provide a fully satisfactory explanation but I would make the following comments:

i) *Darwin's personality*

The fact that Darwin maintained his agnosticism in his letters does not necessarily prove that, privately, as he approached his death, he did not seek for that word on the after-life, which only a religious faith can give.

As already mentioned regarding Darwin's 'private' autobiography, he was well aware that what he wrote was likely to be read widely by a large number of people, both critics and supporters. He would therefore be keen to maintain his 'scientific' image when writing to his evolutionary colleagues. His private thoughts on religion, however, for long suppressed since his early years, may have nevertheless come to the surface. That his published letters to his scientific friends made no reference to any religious views does not necessarily invalidate the authenticity of Lady Hope's account. Indeed, it would be characteristic of his vacillating personality, such as we have seen prior to the publication of his *Origins*, that he would wish to have the reassurance of the Christian faith yet *at the same time* retain the public adulation which he had received due to his work on evolution. In addition, his evasive excuse that he was "a young man with unformed ideas" which "others" made a religion of, is equally characteristic of his natural desire to escape responsibility for the results of his theory — results which he now perceived with greater clarity.

ii) *Fabrication*

It is difficult to imagine a deeply religious Christian, whoever they may be, fabricating the account and then publishing it. Any check which revealed its falsity would bring great discredit upon the originator. Similarly, the suggestion that Emma Wedgwood had a hand in such a story is unacceptable.

Rusch's failure to find the account in the Boston *Watchman Examiner* is certainly an obstacle. However, it should be noted that we do not know the date when the account was printed, whilst Rusch says he examined all the copies available. It may well have appeared in one of those he was unable to find.

There is one slim piece of evidence regarding the printing of the story. The account which appeared in the *Bombay Guardian* in 1916 is referred to by Professor Enoch in his book *Evolution or Creation?* in which he gives the precise date of 25th March. As he was of Indian birth and worked in that country, we could presume that he saw it in this paper. This would verify that at least one account was printed. If the Lady Hope we refer to was the one who visited Darwin, then the publishers of the *Bombay Guardian* (or the publication from which they obtained it) would have been unlikely to have fabricated the story in her name as Lady Hope did not die until 1922. This return us to the question with which we started — who wrote the account in the first place?

Summary

On the basis of the evidence available so far, one could not say with any certainty that Darwin did return to the Christian faith. One can only leave it to the reader to draw his own conclusions from the evidence presented. Those who are interested in tracing genealogies may like to investigate the life of Lady Hope further. Were it to be shown that she was evangelical in faith or that she visited the Downe area in 1882, this would very greatly assist in verifying the accuracy of the account.

However, even if it were eventually to be proven that Darwin did return to the Christian faith in his last years, let me hastily add (lest my creationist colleagues raise their 'hurrahs' too soon) that this would have little effect upon the convinced evolutionist. He will most likely simply dismiss it as a weakness of Darwin in his old age. Furthermore it will make absolutely no difference to his 'scientific' outlook. Surrounded as he is by the debris of a theory which has been progressively broken down, the fact that his hero recanted at the last would be just one more addition to the wreckage. He has enshrined the dogma of 'evolution in some form' above criticism and to it he must hold — for he has nowhere else to go.

APPENDIX II

THE NATURAL HISTORY MUSEUM'S NEW EXHIBITIONS

The following are a few comments upon the various special exhibits which have been prepared by the Public Services Department of the British Natural History Museum at Cromwell Rd.

A) HUMAN BIOLOGY

This exhibition, the largest in the museum, deals with the way in which the human body works. A seemingly disproportionately large number — say some 30% of the displays — are about the subject of sexual reproduction. Full details of the whole process from fertilisation to the actual birth are graphically and colourfully described, with the reproductive organs appearing frequently in dislpays. Nothing is left to the imagination. Even when a family group is shown they are often life size and completely naked whilst in the centre of the first room is a section through two bodies in the coitus position. Throughout the whole of this exhibition, man is only considered as a machine composed of various organs which work together in different ways.

B) ECOLOGY

It might be thought that in dealing with this subject that there would be little opportunity for evolutionary propaganda. Such is not the case however, for the main topic is the 'efficiency' of 'top carnivores' who are the most successful predators upon those who are 'lower' in the 'food chains'. The emphasis throughout is on the implied superiority of a 'top carnivore'. One case shows a cartoon figure of a fox gaily enjoying life, drinking wine etc. and bears the caption "A successful top carnivore feeds at the end of many food chains". Another case deals with 'trophic levels'. Plants are at level one, herbivores are at level two and carnivores at level three with top carnivores at level four. One case has the comment "The more energy living things at one trophic level capture from the one below, the more efficient they are said to be" and the top trophic level cartoon 'man' is holding his hands above his head like a victorious conquerer!

I must say that I consider this display one of the subtlest ways of teaching young children by the 'soft sell' technique that as they are only animals themselves, it is perfectly praiseworthy and 'natural' to be a predator upon those who are below you (and therefore weaker) in the 'chain'.

The main purpose of this display seems to be the demonstration that Nature, "red in tooth and claw", is engaged in the "struggle for survival" with the victory going to the strongest — and all presented in the innocency of a 'Nature lesson'.

C) THE ORIGIN OF SPECIES

On the first display, two alternative views are put forward:

"One view is all living things have developed by a process of gradual change over a very long period of time. This is what we mean by Evolution. Another view is that God created all living things perfect and unchanging[?]. He created each one for a special purpose and this is the basis for the doctrine of Creation."

This statement referring to creation as a possible alternative to evolution has aroused a few academics to condemn the exhibition for making such an admission. They have little cause to complain however for the rest of the displays present the case for evolution in the usual biased fashion.

Rather than give a full description of the various displays we will just refer to three which are typical of the approach.

The first is the admitted difficulty of defining a species. The problems are stated and the ways of overcoming them given. These however are all for closely related animals and the problem of how completely new *groups* of animals could have appeared is totally ignored.

The second is a picture of a young cuckoo in a nest with a caption saying:

"About ten hours after hatching, a young cuckoo starts to push the other eggs out of the nest. This behaviour is inherited. It increases the cuckoo's chances of survival by removing other young that would compete with it for food."

I could not help thinking what an appalling example this sets for motivation of the nations youth.

The third is the "Natural Selection Game" which is played on a computerised television screen and shows how different coloured mice survive against varying backgrounds. During the course of pressing the buttons, the players learn that "With each generation the proportion of well adapted individuals increases."

With all the evidence which the display gives in support of the theory, it is a little surprising that the final case makes the rather lame statement:

"Darwin's theory of evolution by natural selection remains credible even today".

D) MAN'S PLACE IN EVOLUTION

I have criticised the fossil evidence in my earlier work [46]. In this it was shown that one of the most important fossil discoveries (1470 Man) has been deliberately omitted, and that 'cladograms' of the relationships between the species obscures the fact that not one of them are shown as being on the direct line of man's ancestry.

This display is also one of the most clear cases of indoctrination of the public with the myth that they are descended from apes. For example, the very first diplay — of life size models of a naked man and woman — bears the very question-begging statement:

"If[!] life on earth appeared only once, then all living things must be descended from a single common ancestor, and all living things must be related to each other".

Similarly, the beautifully illustrated book about the display has as its opening statement:

"Man — Homo sapiens — is only one of many thousands of animal species alive today".

This idea that humans are only descendants of apes is emphasised throughout the display, and particularly in the first three cases by a question and answer routine. The first case gives the main characteristics of mammals and then asks " Are you a mammal?" By pressing the only knob on the display (no choice here!) the answer "Yes. You are a mammal." appears. Similar questions about primates and apes have the answer — "Yes. You are a primate", and "Yes. You are an ape". The young visitor is left in no doubt when he leaves this exhibition regarding his animal ancestors!

The effectiveness of this indoctrination was brought home to me whilst I was on one of my visits to the museum. The display cases were surrounded by a number of young schoolchildren. The master, in order to emphasise the information they had been looking at, was asking them in a parrot-like fashion "What are you ? — yes, you are a mammal. And are you a primate ? Yes you are a primate" and "Yes, you are an ape." I am sure that this master was convinced that he was conscientiously teaching his boys simple biological facts. Yet that same master, when faced by those boys later in their teenage rebellious years will no doubt admonish them by telling them to stop behaving like animals — completely forgetting that he had already instructed them with this 'fact' only a few years previously !

If people (and particularly the young) are told sufficiently frequently that they are *only* an animal, then we have no cause to complain if eventually they act like one !

APPENDIX III

SYNOPSIS—SCIENTIFIC EVIDENCE AGAINST EVOLUTION

The theory of evolution is nowadays taught as a 'proven fact'. There is however a considerable amount of scientific evidence against it which never receives any publicity in the mass media. The following are just *some* of the extensive array of facts which contradict the theory.

1. FOSSILS

A) The oldest rocks (Pre-Cambrian) have been searched for many years but no undisputed fossils have been found. The Cambrian rocks immediately above, however, contain numerous fully developed complex invertebrates. This sudden appearance of life in the strata has been a major problem for the evolutionists. See Fig. 1

   More recently, minute objects found in Precambrian strata are claimed to be primordeal cells. Some of these however have been found to be nothing more than weathered crystals [65]. At the present, the whole subject is somewhat anomalous and confused. In any case, there is an enormous gap between such microscopic objects and the complex invertebrates such as the trilobites which suddenly appear perfectly formed in the Cambrian strata above them.

B) Despite searching the strata for over 100 years, fossils which would close the gaps between classes and even species have *not* been found, as many evolutionists are now prepared to admit. The very damaging admission by Dr. Niles Eldredge of the American Museum of Natural History that these gaps still exist is given in Appendix VII.

2. THE HORSE 'SERIES'

When challenged to produce a series of fossils demonstrating the transition of one species into another, the 4-3-1 toe evolution of the horse is frequently presented as evidence. However,

A) Over twenty different geneological 'trees' have been drawn up by various scientists. This is because there are 250 similar looking animals to chose from. Those which contradict the series are ignored.

B) All the known species of birds and mammals appear and 'diversify' within the last 150 Million years according to the evolutionists geological time scale. At this rate, the 70 million years it has taken simply to modify a horse's hoof is far too large a proportion of the time since mammals first appeared. There is therefore something seriously wrong with the time scale.

C) Some animals used in the sequence have differing numbers of ribs and lumbar vertebrae, indicating that various species have been used to compile the series. Thus whilst attention has been drawn to the supposed 'evolution' of the hoof, the changing number of vertebrae and ribs has been quietly ignored as this contradicts the theory.

D) Fossils of these animals are mostly found in *America*. Yet the first fossils of modern horses they are supposed to lead up to are found in *Europe*. (Present American horses are a recent introduction).

Just how false this much publicised tree of horse evolution is can be judged by the admissions of two evolutionists - George Gaylord Simpson and Charles Deperet - which are given in Appendix VII.

3. ARCHAEOPTERYX

This bird, about the size of a pigeon, is claimed to be the link between reptiles and birds. But it had perfectly formed feathers which are very complex in design. Examination of a feather under a microscope shows that there are minute hooks (barbs and barbules) and there can be over a million of them on one feather. Nothing which is half way between a feather and a reptile's scale has ever been found. An animal with *half* developed wings could neither run properly nor fly and would be quickly eliminated. Finally, Archaeopteryx is irrelevant, as the fossil bones of a normal bird have been found in strata of the same dating as Archeopteryx. [*Science News* 24th Sept. 1977 p198].

4. BIRDS

A) Evolutionists cannot determine how birds evolved by studying existing species. Special types of skulls, feathers, hollow bones, etc., appear 'randomly' throughout existing species making classification impossible.

B) Nesting habits of some birds cannot be learnt, e.g. the mud nest of the House Martin has to be right first time or the eggs will fall, destroying that generation.

5. GENETIC EXPERIMENTS

After breeding over one million fruit flies, they still obstinately remain fruit flies! There is a wide variety of dog *breeds*, but they are still dogs. Species can vary within limits which cannot be exceeded without producing serious deformities.

6. RECAPITULATION THEORY

This is the theory that the development of a fertilised germ cell retraces the history of the species.(e.g. that gill 'slits' in the human embryo are relics of its fish ancestry). This theory, once hailed as the Biogenitc LAW is now discredited even by evolutionists (See Appendix VII). However it is still implied in some books. Prof. Haeckel (a fiery supporter of Darwin) faked his drawings to support the theory but was convicted by a University court.

7. ORIGIN OF LIFE

A) Passing a spark through a mixture of gasses formed simple amino acids but -

a) they are only the very simplest of 'building blocks' used in the formation of larger organicmolecules.

b) they must be caught in a cold trap to prevent the spark from destroying them,

    c) a reducing (non oxygen) atmosphere is necessary.

    d) any amino acids forming would have been destroyed by the ultraviolet rays of the Sun.

i12 These conditions would not have occurred in nature.

B) There has been insufficient time or material in the whole universe for very complex organic molecules to have formed BY CHANCE. For example, a protein may consist of up to 2,000 say amino acids long every one of which has to be arranged in a particular sequence. If we take just a simple protenoid of say only 100 amino acids length, as there are 20 different amino acids, then the possibility of them coming together by pure chance is 20 raised to the power of 100 which is approximately equal to 10 to the power of 130 (one with 130 noughts behind it). To give some idea of just how large this number is, the total number of atoms in the whole of the known universe is estimated as about 10 to the power of 76 !

Thus with such heavy odds against the formation of life by natural causes being enormously high, then the theory of evolution cannot even clear its first hurdle. Clearly, the assembly of such complex units in a precise order shows the hand of a Designer! It is this fact which forced Prof. Hoyle to make the admissions given in Appendix VII.

8. PEPPERED MOTH

There are two *varieties* - the light and the dark. Elimination of the light variety is *not* evolution despite the claims of witnessing 'evolution in action'. The moths are *still* Peppered Moths.

9. WHALES

Evolutionists are unable to explain how the whale, which is a mammal, went back into the sea without leaving any fossil evidence of intermediate forms.

10. DUCK BILLED PLATYPUS

This strange animal has:

    a) a bill and lays eggs like a duck,

    b) fur like an animal,

    c) webbed and clawed feet,

    d) pockets in its jaws to carry food,

    e) a spur on rear legs which is poisonous like a snake's fang.

A question for the evolutionist- what were its ancestors?

11. RADIOMETRIC DATING

This method is used to give an age to rocks (and thereby the fossils they bear) but it rests upon several unproveable assumptions, e.g.

    a) Radioactive conditions are the same today as they were millions of years ago.

    b) The 'half life' of the elements is constant.

c) The products of the radioactive decay were not originally present nor added since the formation of the rock.

When the same stratum is tested by different methods or even by the same method, it frequently gives an enormous range of ages. For example, one rock gave 14, 30, 95 and 750 million years by different methods. In another case, dating of the same rock for Leakey's 1470 'Man' gave 220 million years and 2.6 million years using the Potassium-Argon method. It is sometimes said that, despite discrepancies, radiometric dating shows that rocks are millions of years old, not thousands. The simple reply is that the 'daughter' elements found in some rocks are naturally occurring along with many other elements. It is claimed that the isochron technique effectively eliminates major assumptions and gives reliable results. Yet some results are admitted to be wildly incorrect - indeed at times giving an age greater than that for the Earth itself![70] The whole subject of Radiometric dating is very technical and really requires a separate treatise. To infer vast ages from the ratios of the elements found in rocks is unwarranted. The *only* reason why the results of Radiometric Dating tests are quoted is that they give ages in terms of *millions* of years. Other methods which yield results of only thousands are completely ignored.

(Note - see Addendum p.208)

CARBON 14

This is a radioactive form of Carbon which is generated in the upper atmosphere by cosmic rays, and all living organisms have a small amount of C14 within them. However, the level is not constant as the ground level activity of 1.63 is *still* rising to generation rate in the upper atmosphere of 2.5, i.e. the amount of C14 is not yet in equilibrium. This makes the true age shorter than apparent age (Fig. 2). This method is quite unreliable for ages over 3,000 years, despite datings up to 40,000 years being quoted.

APPENDIX IV

THE APE-MEN FALLACIES

[This appendix is a summary of APE-MEN FACT OR FALLACY? "(2nd Enlarged Edition 1981) by the same author. Sovereign Publications, Box 88, Bromley, Kent, BR2 9PF. £3.80 inc. post & packing].

Enormous efforts have been made to discover the missing links between Man and the apes. The results, however, are a small collection of unconvincing fossil bones.

1. RECONSTRUCTIONS
A) *Artists' impressions*. With each new discovery, invariably an artist🗊 impression is given showing "what our ancestors looked like". In each case, however, drawings by various artists based on the same skull are completely different ,proving that these pictures are figments of the imagination (Figs. 3,4,5 and 6).

B) *Java 'man'* consists of only a (gibbon's) skull cap and a human leg bone. Yet on these a complete face and body have been reconstructed (Fig.5).

C) *Hesperopithecus (Nebraska man)*. One tooth was found in America and was claimed to be a new ape-man. A complete detailed picture was published in the London Illustrated News of the ape-man and his wife. The tooth was later found to be that of an extinct pig! Little publicity was given to *this* fact.

2. HOMO SAPIENS (Modern Man)
Fossils have been found in older layers than those of the so-called ape-men, but these are ignored by evolutionists, being classed as "forgeries" or "intrusive burials".

3. NEANDERTHAL MAN
They were *not* Man's ancestor as Homo sapiens has been found in earlier strata. He was a degenerate variety of Homo Sapiens having a larger brain and suffering from rickets, osteoarthritis and syphilis. He vanished from Europe, being replaced by modern Homo sapiens from the Middle East.

4. PILTDOWN 'MAN'
Now known to be a fraud, Dawson the amateur is usually considered to be the hoaxer. Teilhard de Chardin, however, who helped with the digging is by far the most likely culprit:

A) A (radioactive) tooth was found, which certainly came from Ichkeul in N. Tunisia. Teilhard had visited this site.

B) An elephant's bone was found which probably came either from the Dordogne in France or from Egypt. Teilhard was born near the

Dordogne and was a teacher in Cairo.

C) Chemical staining of the fake jaw was an involved technical affair. Teilhard lectured in Chemistry and Physics at Cairo University.

D) The whole fraud was the work of an expert who could fool other scientists. Teilhard was a renowned authority in anthropology and palaeontology.

E) He found a fake flint and tooth in the first few days of digging, and later found the important canine tooth in spread gravel that had already been searched.

F) He met Dawson in 1909, three years before the fake stained jaw was discovered. He was studying at the Jesuit College at Hastings where he was ordained in 1911. He therefore had ample time to prepare any fakes.

Recent revelations

In a letter to The Times (November 25th 1978), Dr. Halstead (who once worked in the Natural History Museum) claimed that "according to Hinton (a former Keeper of Zoology) the Piltdown man hoax was planned and executed within the Museum" and that others including Teilhard were involved. This is confirmed by the secrecy surrounding the fossil, experts only being allowed closely to examine plaster replicas. This indicates that the hoax was known at a very high level. The hoax was too elaborate for a practical joke and in addition, if it were intended to later reveal the truth, it would have been far more damaging to those who planned it than the one it was intended to fool — said to be Sir Arthur Smith-Woodward. The most likely motive was to provide evidence that man came from apes, as no fossils had been found for twenty years — since Dubois' discovery of "Java man" in 1890.

5. PEKIN 'MAN'

Teilhard de Chardin worked on this site also, with Dr. Davidson Black (who had also visited the site at Piltdown).

A) Almost every skull was broken into small pieces.

B) Only (incomplete) skulls were found, virtually no limb bones (Fig.4). To explain this, the experts said that Pekin man was a head hunting cannibal! It is obvious that they are only the skulls of large monkeys broken open by real men to obtain the brains for cooking.

Professor Breuil visited the site and saw:

a) stone and bone tools of an advanced type in large quantities,

b) a 23 ft. (7 m) high heap of ash. This was referred to as "traces of fire" by the investigators! He wrote a paper describing what he had seen, but it was omitted from a "complete bibliography" compiled by the experts.

In 1934, skulls of modern men were found alongside the site. Dr. Black died of a heart attack whilst examining them. His successor , Dr. Weidenreich, did not publish anything about them for five years. All the fossils were "lost" at the time of Pearl Harbour and cannot now be examined to check the reconstructions.

In December 1923 world wide publicity was given to the discovery of the skeletons of ten men on the Pekin Man site. Dr. Davidson Black promised to make an important announcement about the new discoveries. Nothing more was ever mentioned about them in any scientific periodical. They simply vanished! What could have happened? It seems likely that, after the blaze of publicity, on closer inspection they were found to be human beings and not ape-men missing links and were therefore quietly ignored.

6. JAVA 'MAN'

In 1890 Dubois found a skull cap (of a giant gibbon) and a human leg bone 45 ft. away (Fig. 5). He put them together and said he had found a "walking ape" (Pithecanthropus erectus). For thirty years he kept secret two skulls and other fossils of modern men he had found at the same time. The Selenka Trinil Expedition could find no further trace of Dubois' ape-man. Von Koenigswald found only a few broken skull pieces and parts of jaws which he claimed confirmed Java man's existence (Fig. 6). Four years before he died, Dubois admitted that he had only found the skull cap of a large gibbon but this was ignored by the experts as by this time Java 'Man' was too well established.

7. SOUTH AFRICAN APE-MEN (Australopithecenes)

All these fossil bones are only those of apes. In a symposium, edited by Sir Julian Huxley, Sir Solly Zuckerman completely rejects these fossil apes.[73p347]. Similarly Oxnard claims that they are not ancestral to man.[*Nature* 4th December 1975 p389]

8. EAST AFRICAN FOSSILS

Ramapithecus

The *only* fossils of this ape are some thirty jaws and teeth! (Fig. 3). This species is given much publicity as it is the only possible ancestor of some more recent discoveries-

Richard Leakey

"1470 Man" (Fig. 3): Almost certainly this is a small human skull as it has many human features, and human leg bones have been found in the same strata. First radiometric dating of strata gave the impossibly high figure of 220 million years. This was rejected and a second sample gave 2.6 million years. It was this latter date which received great publicity. The human characteristics of 1470 man are so embarrassing that it has been reclassified as Homo habilis and is now quietly ignored.

D. Johanson

"Lucy" is about 40% of the skeleton of an ape. There is virtually no evidence that this fossil was evolving into Homo sapiens. The remainder of Johanson's discoveries is a meagre collection of apes bones.

Laetolil Footprints

Dated as being 3.6 million years old, a careful reading of the reports [Nature 22nd March 1979 and National Geographic Magazine April

1979] show that these 'hominid' footprints are those of modern men. To admit that they were human footprints would have upset the supposed theory of the evolution of man. Yet human footprints have been found alongside those of long extinct dinosaurs.

APPENDIX V

SYNOPSIS-SCIENTIFIC EVIDENCE
SUPPORTING GENESIS

Genesis chapter 1 v1: "In the beginning God created the heaven and the earth". God created *time*, the heaven and an empty earth.

Pleochroic Haloes

Some radioactive isotopes of Polonium, Bismuth and Lead have half lives of only a few *minutes*. Yet inclusions of them in solidified mica in Pre-Cambrian strata have generated dark rings of decay products with no evidence of earlier products of decaying elements. As rings would not appear in *molten* mica, this indicates that *both* materials were *created* in an instant of time and within minutes the radioactivity had darkened the mica crystals. This supports the instant *creation* of the earth, as given in Gen.1 v1. When the evidence was submitted to a number of top experts around the world, they were unable to fault the facts despite repeating the tests and were subsequently very guarded in commenting on the obvious implications.

1 v5: "The evening and the morning were the first day". Emphatic repetition reinforces the fact that these were twenty four hour periods, *not* thousands or millions of years.

1 v6: Dividing of the waters. A gaseous transparent water *vapour* canopy was placed around the earth (Fig. 7). This had the following results:
   a) Lethal cosmic rays were blocked from reaching the earth's surface.
   b) Heat from the sun would be dispersed uniformly. With a uniform temperature, there would be no high winds.
   c) No clouds or rain — watering of ground was by "streams from the earth".[2v6]
   d) Worldwide lush vegetation. This would explain the presence of coal found at South Pole and Spitzbergen.
Scientists cannot account for the amount of He3 in the atmosphere. Prof. Korff has said that it may have been "possibly generated by neutrons when it [the atmosphere] contained much more water vapour"!

1 v11: Sudden creation
   a) Numerous invertebrates appear suddenly in rock strata above Pre-Cambrian devoid of complex fossils. All species appear perfect, not half developed (Fig.   ).

b) "After his kind". Species can vary but only within limits. They do *not* evolve or mutate into other species.

1v14-19: *Creation of Sun, Moon, stars, etc.*

No satisfactory explanation has ever been provided of how the planetary system came into existence:

a) Each planet has the right speed for its distance from the sun to maintain a roughly circular orbit (but Pluto swings inside Neptune's orbit).

b) Origin of planets' and moons' angular momentum is unknown, as Sun has very little angular momentum.

c) Planets have widely different compositions.

d) Venus is slowly spinning in the *opposite* direction to its rotation round the sun, whilst Uranus' axis of spin is almost "horizontal".

e) Of the thirty three major moons circling the planets, eleven of them rotate in the opposite direction to the planet's rotation around the sun. Jupiter and Saturn have moons going in *both* directions. Uranus, originally said to have five moons, all retrograde, now found to have a hundred moons!

f) Spiral arms of galactic nebula should have closed up in one-tenth of time of earth's age (which is supposed to be 4,500 million years).

Young age of earth

A) Micro-meteoric dust is descending on to this planet at a rate of 14 million tons each year:

a) If the earth is 4,500 million years old, then the depth of dust should be about 54 ft thick. There is approximately one-eighth of an inch depth on Moon's surface. This gives an age in the order of only 10,000 years.

b) The dust contains a percentage of Nickel, and if it is assumed that most of the dust is swept into the sea then tests show that there is only 8,000 years' worth of Nickel in the oceans.

c) Sun's gravitation would have swept the whole planetary system clear of dust within 10,000 years.

B) Moon — is still warm, has a magnetic field and suffers from moonquakes - all indicating a young age.

C) Comets — lose matter as they go round the sun, and short term comets should have disappeared in 10,000 years or so.

D) Helium in atmosphere — even if it *all* came from decay of Uranium, Thorium and cosmic rays there is less than 12,000 years' worth.

E) Sun's gravitational collapse. It has been found that the sun is shrinking in the order of 0.01% per century (originally thought to be as much as 0.1% [71]).

a) The energy generated by this collapse is more than sufficient to provide the enormous heat radiated by the sun.

b) This being so, the source of the suns power (and that of other stars!) is *not* due to thermonuclear energy. This is confirmed by the failure of scientists to detect emissions which would prove that radio-activity was its source of power.

c) If it has been shrinking at the rate of 0.01% per century, then only about 1 million years ago its diameter would have been equal to the earths orbit.

v16: "He made the stars also"
Consider the amount of energy in just *one* atom.

v26: Creation of Man.
There are *no* links between Man and the animals. So-called ape-men do not bear close examination. (See Appendix III)

Gen.2 v4: Second account of creation centred on Man.

6v4: "Giants in the earth".
Eighteen inch long human footprints have been found in Cretaceous solidified mud in the bed of the Paluxy River in Texas, along with normal footprints and dinosaur prints. Yet dinosaurs are supposed to have died out 80 million years before Man!

7v11: Noah's Flood. "Fountains of deep broken and windows of heaven opened"
Collapse of Water Vapour Canopy which would produce torrential rain.

Existing Strata
A) Animals are often found entombed in an attitude of fear - head back, mouth open, fins or wings extended, spines erect, etc.
B) Fossils sometimes found with scales and skin still visible, having had no time to decay. Fossilisation requires rapid inundation and fossils are not being formed under todays conditions. Bones just simply disintegrate and disappear.
C) Fossil sequence in geological strata show:
   a) *Invertebrates* (slow moving marine animals) would perish first followed by the more mobile fishes who would in turn be overwhelmed by the silt brought down by the flood waters.
   b) *Amphibia* (who would be close to the edge of the sea) would perish next as the waters rose.
   c) *Reptiles* (slow moving land animals) would be the next to die.
   d) *Mammals* could flee from rising water, the larger, faster ones surviving the longest.
   e) *Man* would exercise most ingenuity — clinging to logs, etc. to escape the flood.
The sequence outlined above is a perfectly satisfactory explanation of the order in which the various fossils are found in the strata. It is *not* the order in which they evolved over millions of years but the order in which they were inundated at the time of Noah's flood. There are many 'anomalies' in the geological sequence (such as dinosaur tracks and human footprints in the same strata) which are far better explained by the creationist model than the theory of evolution.

9v13: "I set my bow to the clouds".

The atmosphere now has clouds and raindrops and the rainbow is now visible for the first time.

Age of Patriarchs

Up to Noah, all the patriarchs lived just under 1000 years (Fig. 8). After the flood, the ages decline roughly exponentially to a new low level of 70 years. This is due to the collapse of the Water Vapour Canopy which now exposed the earth to lethal cosmic rays. These damage genetic material in the cells of all living organisms, resulting in their earlier death. (Animals subjected to low levels of radiation have shorter lives. Hospital radiologists have shorter life expectancy.)

Flood Stories

Numerous tribes have a legend of a great flood in ancient times, and the more remote from the Middle East, the more corrupt the version. The experts in their effort to discredit the Bible, claim that the Genesis account of the Flood is a copy from the Babylonian Gilgamesh Epic. The real truth is quite the opposite. The Gilgamesh epic is a corrupt mythical pagan version of the event of the worldwide flood, of which the Genesis account is an accurate record.

Carbon 14

See Section 1. Calculated rise of C14 activity began less than 9000 years ago, probably when the Water Vapour Canopy collapsed. (See Fig. 2)

Addendum – Decay in the speed of light

Having largely completed the typesetting of the text, two very important articles were read in *Ex Nihilo* (vol.4 No's 1 and 3 1981) – the periodical of the Australian *Creation Science Publishing* (see Appendix VI). In these articles, the author, Barry Setterfield, shows conclusively that the speed of light – always considered to be a fixed value by scientists – has actually decreased since it was first measured in 1675.

There is insufficient space in a work of this nature to treat the whole subject with the attention which it deserves. However, so important are its implications for creationists that a brief summary is given here. A synopsis of these papers and a third as yet unpublished will appear in pamphlet No. 230 of the Creation Science Movement. For full details however the original papers should be consulted.

The 1675 measurement is generally acknowledged to be quite accurate. Subsequent measurements, however, show a decline to about 1960 after which no significant change could be detected. After numerous attempts, a curve was found to fit the results exactly – a log sine curve. Having thus established the law governing the decay of light, it could now be accurately extrapolated backwards in time to ascertain its velocity in past ages.

Setterfield shows that whilst some atomic relationships vary with the speed of light (both in his theory and by measured results) others do not, and all chemical and molecular laws remain unaffected. In addition, in his third article, Setterfield promises to show how the change in the speed of light is due to a *collapsing* universe.

These papers have been criticized, but Setterfield gives very satisfactory replies to the objections raised. If these results are accepted, they provide very considerable support to creationists on the following points.

1. *Radiometric decay*

Radioactivity is proportional to the speed of light – in which case the decay of radioactive material in the rocks in past ages would be far higher than it is today. *This decay would be so rapid that all ages could be accommodated within a six thousand year period.* Thus the reason for the apparent great age of rocks is easily explained by the high decay rates in the past.

2. *Light from distant galaxies*

An excuse often given for rejecting Biblical chronology is that light we now see from distant galaxies has been travelling through space for vast periods of time. This objection is now easily solved, for at the time of Creation light would have been at almost infinite speed. Therefore light from even the most distant galaxy would have been visible almost instantaneously everywhere in the Universe. It is interesting that this also solves the problem of objects in galaxies which appear to be separating faster than the speed of light (as measured today).

3. *Start of decay*

It is this aspect which produces one of the most surprising results. The log sine curve fitted the experimental data exactly. This curve is particularly sensitive to the precise starting date when the speed began to decay. From the shape of the curve, there is a limit on the time in the past from which it could have begun decaying, i.e. the curve is almost vertical to a particular date in the past. When this limiting date is calculated, it comes to the surprising date of 4040 B.C. ±20 years.

It must be remembered that this conclusion is based entirely upon experimental data without any reference to Biblical chronology whatsoever. As is well known, several Bible scholars have calculated the date for the Creation/Fall and have arrived at dates close to 4000 B.C. – Archbishop Usher's date of 4004 is the one most usually referred to. The fact that this date has now received scientific confirmation should greatly assist the Creationist cause. No doubt, evolutionists will proffer various explanations, but they will be hypotheses and not based upon the observed facts.

Religeous implications

This striking confirmation of the accuracy of the Bible cannot but make one ponder on its theological implications. For many years – ever since the Uniformitarian Theory dominated scientific thinking – one of the most frequent excuses for rejecting the accuracy of the Bible has been such 'facts' as the time taken for light to reach us from distant galaxies, radiometric dates and similar 'scientific evidence'. Yet all these objections are very simply answered by a decrease in the speed of light – a clear demonstration of the foolishness of relying on (pseudo-)'science' rather than the truth of God's Word.

On that Great Day when each man will be held responsible for his deliberate rejection of God's forgiveness for his rebellion, every possible excuse he makes will be shown to be inadequate. Not only will he be judged accordingly, but he will also suffer the remorse of acknowledging that his judgement is fully deserved for he had deliberately ignored the many evidences of God's hand in His Creation.

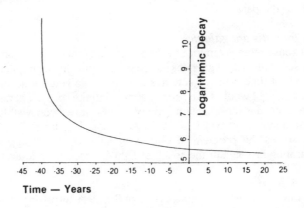

Fig.9. Decay of the speed of light

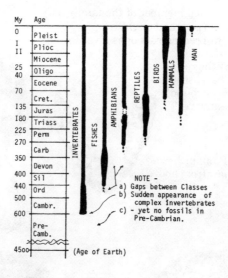

Fig.1. The evolutionary view of the geological strata

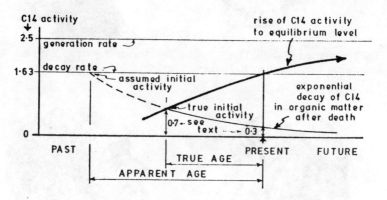

CARBON I4 dating errors. Sample with
activity of ·3 would have started with ·7,
not with I·63 assumed. Level of activity
probably started to rise at time of FLOOD.

Fig.2. Carbon 14 dating

RAMAPITHECUS

This much publicised member of the family only consists of some 30 jaws and teeth!
Nevertheless a full reconstruction is given in "Origins" on p.67.
The authors say the reconstruction " . . . because so few remains have been found, must be very tentative"!

Ramapithecus
jaw from above

1470 "MAN"

The reconstructed skull. The fragments were found by R. Leakey in 1972.

As drawn for the National Geographical Magazine.

Sunday Times
Nov 12th, 1977

Observer
Nov. 12th, 1977

Fig.3. Some reconstructions

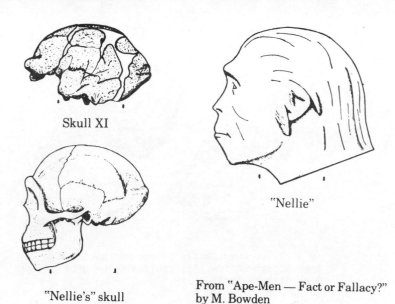

Skull XI

"Nellie"

"Nellie's" skull

From "Ape-Men — Fact or Fallacy?"
by M. Bowden

Fig.4. Pekin 'Man'

This famous 'link' consists only of broken pieces of (apes) skull bones and jaws. With the help of much plaster, five skulls were assembled the most complete of which was skull XI. To this was added a jaw found 80 feet higher in the excavation, to make a complete skull, on which was modelled a womans face — "Nellie".

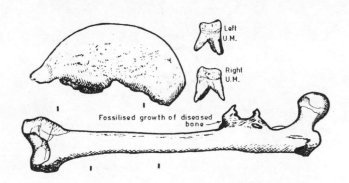

Left
U.M.

Right
U.M.

Fossilised growth of diseased bone

The bones upon which Dubois constructed his JAVA "MAN"

Fig.5. Java 'Man'

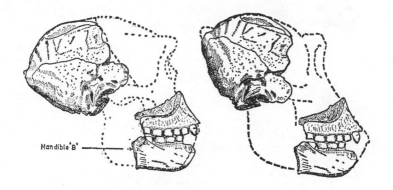

Von Koenigswald's Java man bones on
the reconstruction (A), could also be made to
look like a gorilla ! (B)

Fig.6. Von Koenigswald's reconstruction

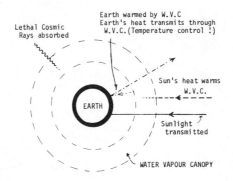

Fig.7. The Water Vapour Canopy

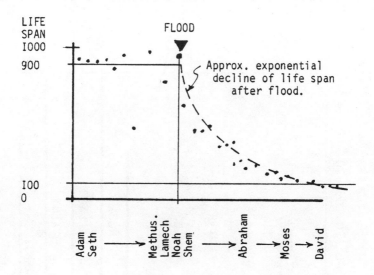

Fig.8. The ages of the Patriarchs

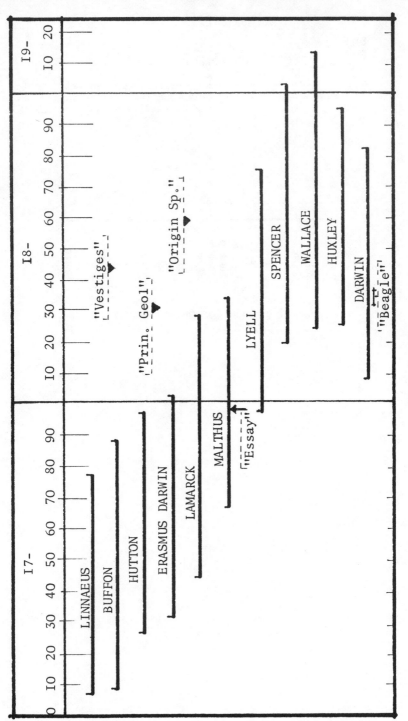

Time Scale of the Rise of Evolution.

APPENDIX VI

CREATIONIST ORGANISATIONS

It is to be hoped that many will be prepared either to support one or more of the creationist organisations or may wish to receive up-to-date publications dealing with various aspects of the current debate. Accordingly this appendix gives the main U.K. groups and three overseas organisations.

THE CREATION SCIENCE MOVEMENT

Founded in 1932 as the Evolution Protest Movement, it was the first organisation ever formed to counteract the prevailing dominant evolution theory.

Meetings Usually third Saturday in September at Caxton Hall, Caxton St, London. 2 lectures and a film.

Publications. Scientific pamphlets and book reviews, comments etc.-2 issues per annum.

Membership 800. *Fee* £2

Secretary Mr. B. Hughes, "Rivendell", 20 Foxley Lane, High Salvington, Worthing, West Sussex, BN13 3AB.

THE NEWTON SCIENTIFIC ASSOCIATION

Formed in 1974 by members of the Metropolitan Tabernacle, Elephant and Castle. Main activity is the presentation of the scientific evidence against evolution to those who are evolutionists or are uncommitted. It does not generally deal with a Creation model based on Biblical revelation, but all executives are committed Creationists.

Meetings Two each year at the Metropolitan Tabernacle on Saturday afternoons - the 1st Saturday in March and 3rd Saturday in November. Three lectures at each.

Publications Occasional publications and newsletter. Some talks appear in the Church magazine *The Sword and the Trowel*

Membership 300. *Fee* £2.50 p.a.

Secretary Mr. E. Powell, 77 Oakhampton Rd., Mill Hill, London NW7 1NG.

BIBLICAL CREATION SOCIETY

Formed in 1976 by a group of creationists who were particularly keen to promote the creationist cause amongst undergraduates and staff in universties and theological colleges, teachers, ministers etc at a popular level. As well as the scientific evidence, it deals extensively with the Biblical exposition of Creation and allied topics.

Meetings One annual conference in November at the Westminster Chapel, Buckingham Gate, London. 3 lectures, films etc. Also 3-4 regional conferences and a 3-4 day study conference

Publications-Biblical Creation issued three times each year. A 30-40 page booklet on a variety of creationist topics. Also *Rainbow* - 3/year, 4-8pp, a popular treatment. Pamphlets, Audio-visual materials etc.

Membership 700. *Fee* £5 p.a.(incl *Biblical Creation*)

Secretary Dr. C. Darnbrough, Biblical Creation Society, 51 Cloan Crescent, Bishopbriggs, Glasgow G64 2HN

CREATION NEWS

This is a 4 page broadsheet issued about four times each year which is compiled and circulated by Dr. A.J. Monty White. The contents consist of general information and topics of interest to Creationists. There is no subscription and the leaflets are sent free to anyone wishing to receive them but donations to cover expenses may be given.

Address Dr. A.J. Monty White, 3 Church Terrace, CARDIFF CF2 5AW.

AMERICAN GROUPS

The following two groups are the major organisations in America. They provide excellent publications giving the latest information in the creation /evolution debate.

INSTITUTE OF CREATION RESEARCH

The major American organisation, which has several travelling lecturers, scientific advisers, lecture courses, etc.

Publications "Acts and Facts" - a broadsheet issued monthly with up-to-date information and general topics which is sent free of charge on request.

There is no membership as such, support coming from freewill contributions.

Address 2100 Greenfield Drive, P.O.Box 2666, El Cajon, California 92021, USA.

CREATION RESEARCH SOCIETY

The members of this society must be qualified in science to the level of a Masters Degree at least.

Publications "CRS Quarterly". Each issue has up to 100 pages of excellent scientific articles which are a major source of information for creationists the world over.

Membership Total 700. Associate membership 2000. Fee - $15/yr.

Address Creation Research Society, 2717 Cranbrook Rd., Ann Arbor, Michigan 48104, USA.

AUSTRALIA

CREATION SCIENCE FOUNDATION

A mission team of 5 full time workers which distributes literature and organises lectures. Publishes "Ex Nihilo" (50-80pp) which covers both elementary and technical fields of creationism.

No memberership as such. For subscription for "Ex Nihilo" write to-Creation Science, P. O. Box 302, Sunnybank, 4109.

APPENDIX VII

"QUOTABLE QUOTES" FOR CREATIONISTS

It does not take much reading of evolutionist articles to realize that, whilst many of the writers vehemently promote the evolutionist cause, not infrequently, tucked away in some of the papers (often a technical one) is an admission that in the particular subject being dealt with there is really *no* support for the theory at all - in fact the evidence is against it. Indeed it would be possible to disprove evolution simply by quoting the conclusions of evolutionists when they are referring to their own specialist subjects.

The contents of this appendix were compiled by Mr. J.V. Collyer and the author and first appeared as pamphlet No.228 of the Creation Science Movement in January 1981. We will be considering some of the more notable utterances of evolutionists, non committed scientists and creationists on a variety of topics.

First, however, we will begin with some of the more dogmatic statements made by those who are fanatical supporters of evolution.

EVOLUTIONARY DOGMA

"That evolution, so stated, is an indisputable fact is accepted by all but one or two of those who are accredited experts in the study of biology.... Of the *fact* of organic evolution there can at the present day be no reasonable doubt; the evidences for it are so overwhelming that those who reject it can only be the victims of ignorance or of prejudice." [M.J. Kenny *Teach yourself Evolution* 1966 pp1 & 159]

"Darwin ... finally and definitely established evolution as a fact." [Professor George Gaylord Simpson]

"A belief in Evolution is a basal doctrine in the Rationalists' Liturgy." [Sir Arthur Keith - *Darwinism and its Critics* 1935 p53]

Those who did not honour Darwin "inevitably attract the speculative psychiatric eye to himself." [Garret Hardin of the Californian Institute of Technology *Nature and Man's Fate* Mentor 1961]

"Only ignorance, neglect of truth, or prejudice could be the excuse for those who in the present state of knowledge without discovering new facts in the laboratory or in the field, seek to impugn the scientific evidence for evolution." [Sir Gavin de Beer *A Handbook of Evolution* British Museum (Natural History) 2nd Ed.(1958)]

"Will Darwin's victory endure for all time? .... so strong has his position become that I am convinced that it never can be shaken." [Sir Arthur Keith *Concerning Man's Origin* Watts & Co. 1927]

"Darwin's first point, that man is the product of evolution involving natural selection, has been attacked on emotional grounds, but it was not and is not now honestly questionable on strictly scientific grounds by anyone really familiar with the facts." [G. Gaylord Simpson *The*

Biological Nature of Man Science N.Y. 152:472-478]

By far the most dogmatic evolutionist is Sir Julian Huxley (grandson of Thomas Henry Huxley, the close friend of Darwin who acted as Darwin's "bulldog").

"The first point to make about Darwin's theory is that it is no longer a theory but a fact. No serious scientist would deny the fact that evolution has occurred, just as he would not deny the earth goes round the sun.... all scientists agree that evolution is a fact ... there is absolutely no disagreement." [*Issues in Evolution* vol 3 of *Evolution after Darwin* Sol Tax Editor, Chicago University Press 1960]

"Today, a century after the publication of the *Origin*, Darwin's great discovery, the universal principle of natural selection, is firmly and finally established as the sole agency of major evolutionary change." [Introduction to the Mentor edition of *The Origin of Species* Mentor N.Y.]

ADMITTED WEAKNESSES

Despite such confident statements, many evolutionists have admitted that the case is far from adequately proven. Who better to start with than Darwin himself:

"Not one change of species into another is on record ... we cannot prove that a single species has been changed." [Charles Darwin *My life and letters*]

"To suppose that the eye with all its inimitable contrivances for adjusting the focus to different distances, for admitting different amounts of light, and for the correction of spherical and chromatic aberration, could have been formed by natural selection, seems, I freely confess, absurd in the highest degree." [Charles Darwin *Origin of Species* Chapter "Difficulties"]

"It is good to keep in mind ... that nobody has ever succeeded in producing even one new species by the accumulation of micromutations. Darwin's theory of natural selection has never had any proof, yet it has been universally accepted." [Prof. R. Goldschmidt (PhD., DSc., Prof. Zoology, Univ. Calif. in *Material Basis of Evolution* Yale Un. Press.]

"The facts fail to give any information regarding the origin of actual species, not to mention the higher categories" [Prof. R. Goldschmidt *The Natural Basis of Evolution* p165].

"... by the proper standards of scientific argument, evolution is *not* a fact but a theory powerfully sustained by a wide range of circumstantial evidence and therefore widely accepted. To pretend otherwise does a disservice to science in general and biology in particular." [R. D'Oyley Good "Natural Selection Re-examined" in *The Listener* 61:797-799 and a letter replying to criticism p986]

"It is therefore a matter of faith on the part of the biologist that biogenesis did occur and he can choose whatever method of biogenesis happens to suit him personally; the evidence for what did happen is not available." [Prof. G.A. Kerkut *Implications of Evolution* Pergamon N.Y. 1960 p150]

"The one systematic effect of mutations seems to be a tendency towards degeneration." [Dr. Sewall Wright *The New Systematics* Clarendon Press p174]

"If one allows the unquestionably largest experimenter to speak, namely nature, one gets a clear and incontrovertible answer to the question about the significance of mutations for the formation of species and evolution. They disappear under the competitive conditions of natural selection, as soap bubbles burst in a breeze." [Dr. Heribert Nilsson *Synthetische Artbildung* Gleerup Press p174]

"Palaeontological knowledge regarding man's past history is still of the most fragmentary kind. Each additional scrap becomes the subject of a voluminous literature and the basis of an edifice of speculation out of all proportion to the foundation upon which it rests." [Sir John Graham Kerr, Glasgow Univ. *Evolution* Macmillan 1926 p212]

"The pathetic thing is that we have scientists who are trying to prove evolution, which no scientist can ever prove." [Dr. Robert A. Millikan (Nobel prize winner and eminent evolutionist)]

"The scientists religious feeling takes the form of a rapturous amazement at the harmony of natural law, which reveals an intelligence of such superiority that compared with it, all the systematic thinking and acting of human beings is an utterly insignificant reflection." [Prof. Einstein *The World as I see it* p29]

"The molecule-to-cell transition is a jump of fantastic dimension, which lies beyond the range of testable hypothesis. In this area all is conjecture. The avilable facts do not provide a basis for postulating that cells arose on this planet." [D.E Green and R.F. Goldberger *Molecular Insights into the living process* Academic Press N.Y. 1967 p407]

On the "evolution" of the horse, two experts have admitted:

"The supposed pedigree of the horse is a deceitful delusion, which... in no way enlightens us as to the palaeontological origins of the horse." [Charles Deperet, a French palaeontologist in *Transformations of the Animal World* p105]

"The uniform continuous transformation of Hyracotherium into Equus, so dear to the hearts of generations of text book writers, never happened in nature." [George Gaylord Simpson *Life of the Past* p119].

"A scientific study of the universe has suggested a conclusion which may be summed up... in the statement that the universe appears to have been designed by a pure mathematician" [Sir James Jeans *The Mysterious Universe* p140].

"Modern medicine and surgery are founded on the truth enunciated by Pasteur, that life proceeds only from life, and only from life of the same type and kind" [Dr. McNair Wilson, (Editor *Oxford Medical Publications*) *The Witness of Science* Murray 1942].

"Increase of knowledge about biology has tended to emphasize the extreme rigidity of type, and more and more to discount the idea of transmutation from one type to another - the essential basis of Darwinism" [Dr. McNair Wilson - Ibid]

"The record of the rocks is decidedly against evolutionists" [Sir William Dawson, famous Canadian geologist].

"The theory of evolution suffers from grave defects, which are more and more apparent as time advances. It can no longer square with practical scientific knowledge" [Dr. Albert Fleishmann, Zoologist of Erlangen Univ.].

"To the unprejudiced, the fossil record of plants is in favour of special creation" [Prof. E.J.H. Corner, Camb. Univ. *Contemporary Botanical Thought* p97].

"There is not the slightest evidence that any of the major groups arose from any other" [Dr. Austin Clark F.R.G.S., *Quarterly Review of Biology* Dec. 28 p539].

One notable comment made recently is:

"I think however that we must go further than this and admit that the only acceptable explanation is *Creation* . I know that this is anathema to physicists, as indeed it is to me, but we must not reject a theory that we do not like if the experimental evidence supports it" [Prof. H.J. Lipson F.R.S. "A physicist looks at evolution", *Physics Bulletin* 31 1980 p138].

It is an interesting spectacle to watch two rival views of evolution pointing out the weaknesses of each other. Such is happening at the moment regarding neo-Darwinian 'gradualism' (progress by small mutations) versus 'punctuated equilibrium' (periods of tranquility with abrupt local changes producing completely new species). This latter theory uses 'Cladograms' which show 'relationships' but *not* ancestors, as existing species are thought to have rapidly appeared from earlier species. Those who hold this view, such as Professor S.J. Gould and Dr. Niles Eldredge, have been called 'Marxist' as the theory is more in line with the 'sudden jumps' of revolution in society. Dr. L.B. Halstead has similarly criticised the British Natural History Museum for its extensive use of Cladograms in its new exhibitions. [*Nature* 26th October 1978 v275 p683]

To support the case for 'punctuated equilibrium', Dr. Eldredge has appealed to the gaps in the fossil record [which Creationists have pointed to for many years!]. A synopsis of a talk he gave appeared in the *Guardian* [21st November 1978], which quotes him as saying:

"...the smooth transition from one form of life to another which is implied in the theory is... not borne out by the facts. The search for 'missing links' between various living creatures, like humans and apes, is probably fruitless... because they probably never existed as distinct transitional types... But no one has yet found any evidence of such transitional creatures. This oddity has been attributed to gaps in the fossil record which gradualists expected to fill when rock strata of the proper age had been found. In the last decade, however, geologists have found rock layers of all divisions of the last 500 million years and no transitional forms were contained in them. If it is not the fossil record which is incomplete then it must be the theory"

The *Guardian* then adds "What is extraordinary is that in the 120 years since Darwin appeared to have cracked the problem with elegant neatness in *The Origin of Species*, the principle has withstood all attacks on it - and yet still evolves loose ends."

One of the most unexpected "turn-rounds" of a scientist is that of Prof. Fred Hoyle. For many years he had accepted evolution but he began to study the possibilities of how life could have started on this planet. He found that the chances of it occurring by accident were so unlikely that it was ridiculous to believe that it did happen. He, together with Prof. Wickramasinghe, have written a book setting out their views that life must have come from outer space. Some notable comments in their book *Evolution from Space* [Dent 1981] are:

"Biochemical systems are exceedingly complex, so much so that the chance of their being formed through random shufflings of simple organic molecules is exceedingly minute, to a point indeed where it is insensibly different from zero" [p3].

"The obvious escape route is to look outside the earth, although we should be warned that even this route may not be easy to follow. There is no way in which we can expect to avoid the need for information, no way in which we can simply get by with a bigger and better organic soup, as we ourselves hoped might be possible a year or two ago. The numbers we calculated above are essentially just as unfaceable for a universal soup as a terrestrial one" [p31].

They finish up by claiming that the evidence shows that life must have been designed by a superior intelligence - namely God. The God they are referring to, however, is still a long way removed from the God of love and justice of the Christian faith as we find is clearly set out in the Bible.

CREATIONIST'S COMMENTS

The following are some criticisms of evolution made by some well known creationists.

"It is Darwin's habit of confusing the provable with the unprovable which constituted to my mind, his unforgivable offence against science" [Dr. L.M. Davies *The Bible and Modern Science* Constable 1953 p8].

"The theory of the transmutation of species is a scientific mistake, untrue in its facts, unscientific in its method, and mischievous in its tendency" [Prof. J.L.R. Agassiz,1807-1873, (a famous Harvard Professor who strongly opposed evolution) in *Methods of Study in Natural History*]

"I am not satisfied that Darwin proved his point or that his influence in scientific and public thinking has been beneficial...the success of Darwinism was accomplished by a decline in scientific integrity" [Dr. W.R. Thompson, Ref 59]

"Was the eye contrived without skill in optics, and the ear without knowledge of sounds?" [Sir Isaac Newton *Opticks* New York 1952 pp369-70].

"Evolution is baseless and quite incredible" [Dr. Ambrose Fleming, Pres: Brit. Assoc. Advancement of Science, in *The Unleashing of Evolutionary Thought*].

"If complex organisms ever did evolve from simpler ones, the process took place contrary to the laws of nature, and must have involved what may rightly be termed the miraculous" [Dr. R.E.D. Clark, Victoria Institute 1943 p63].

And finally, some strong words by a most eminent scientist:

"Overwhelming strong proofs of intelligent and benevolent design lie around us...The atheistic idea is so nonsensical that I cannot put it into words" [Lord Kelvin Vict. Inst. No.124 p267].

APPENDIX VIII

FAMOUS CREATIONIST SCIENTISTS

Sometimes the statement is made to the effect that "without a theory of evolution scientists have no framework or reference within which they can work". Another similar claim is that "(blind!) acceptance of the Bible inhibits scientific research as God is called in to explain the unknown".

The following list of famous scientists who were Bible-believing Christians appeared in the January 1981 *Acts and Facts* leaflet issued by the Insitute of Creation Research of America. It gives the fundamental scientific disciplines which they established together with some of the important discoveries they made. To refute the two criticisms given above it requires only its presentation without further comment.

SCIENTIST	SCIENCE (and discoveries)
Agassiz, Louis (1807-1873)	Glacial Geology
	Ichthyology
Babbage, Charles (1792-1871)	Computer Science
	(Actuarial tables)
	(Calculating machine)
Boyle, Robert (1627-1691)	Chemistry
	Gas Dynamics
Brewster, David (1781-1868)	Optical Mineralogy
	(Kaleidoscope)
Cuvier, Georges (1769-1832)	Comparative Anatomy
	Vertebrate Paleontology
Da Vinci, Leonardo (1452-1519)	Hydraulics
Davy, Humphrey (1769-1832)	Thermokinetics
Fabre, Henri (1823-1915)	Entomology of Insects
Faraday, Michael (1791-1867)	Electro-Magnetics
	Field Theory
	(Electric generators)
Fleming, John Ambrose (1849-1945)	Electronics
	(Thermionic valve)
Herschel, William (1738-1822)	Galactic Astronomy
	(Double stars)
Joule, James (1818-1889)	Reversible Thermodynamics
Kelvin, Lord (1824-1907)	Energetics
	Thermodynamics
	(Absolute temperature scale)
	(Transatlantic cable)
Kepler, Johann (1571-1630)	Celestial Mechanics
	Physical Astronomy (Ephemeris tables)
Linnaeus, Carolus (1707-1778)	Systematic Biology
	(Classification system)
Lister, Joseph (1827-1912)	Antiseptic Surgery
Maury, Matthew (1806-1873)	Hydrography

219

SCIENTIST	SCIENCE (and discoveries)
Maxwell, James Clerk (1831-1879)	Electrodynamics
	Statistical Thermodynamics
Mendel, Gregor (1822-1884)	Genetics
Newton, Isaac (1642-1727)	Calculus
	Dynamics (Law of gravity)
Pascal, Blaise (1623-1662)	Hydrostatics
	(Barometer)
Pasteur, Louis (1822-1895)	Bacteriology
(Biogenesis law) (Fermentation control)	
	(Immunisation)
Ramsay, William (1852-1916)	Isotopic Chemistry
	(Inert gases)
Ray, John (1627-1705)	Natural History
Rayleigh, Lord (1842-1919)	Dimensional Analysis
	Model Analysis
Riemann, Bernhard (1826-1866)	Non-Euclidean Geometry
Simpson, James (1811-1870)	Gynecology (Chloroform)
Steno, Nicholas (1631-1686)	Stratigraphy
Stokes, George (1819-1903)	Fluid Mechanics
Virchow, Rudolph (1821-1902)	Pathology
Woodward, John (1665-1728)	Paleontology

APPENDIX IX

BIBLE QUOTATIONS

See to it that no-one takes you captive through hollow and deceptive philosophy, which depends on human tradition and the basic principles of this world rather than on Christ.[Colossians 2 v8]

Timothy, guard what has been entrusted to your care. Turn away from godless chatter and the opposing ideas of what is falsely called knowledge, which some have professed and in so doing have wandered from the faith.[1 Timothy 6 v20-21]

The wrath of God is being revealed from heaven against all the godlessness and wickedness of men who suppress the truth by their wickedness, since what may be known about God is plain to them, because God has made it plain to them. For since the creation of the world God's invisible qualities - his eternal power and divine nature - have been clearly seen, being understood from what has been made, so that men are without any excuse. For although they knew God, they neither glorified him as God nor gave thanks to him, but their thinking became futile and their foolish hearts were darkened. Although they claimed to be wise, they became fools and exchanged the glory of the immortal God for images made to look like mortal man and birds and animals and reptiles.[Romans 1 v18-23]

For the time will come when men will not put up with sound doctrine. Instead, to suit their own desires, they will gather around them a great number of teachers to say what their itching ears want to hear. They will turn their ears away from the truth and turn aside to myths.[2 Timothy 4 v3-4]

My purpose is that they may be encouraged in heart and united in love, so that they may have the full riches of complete understanding, in order that they may know the mystery of God, namely, Christ, in whom are hidden all the treasures of wisdom and knowledge. I tell you this so that no-one may deceive you by fine-sounding arguments.[Colossians 2 v2-4]

But there were also false prophets among the people, just as there will be false teachers among you. They will secretly introduce destructive heresies, even denying the sovereign Lord who bought them - bringing swift destruction on themselves. Many will follow their shameful ways and will bring the way of truth into disrepute. In their greed these teachers will exploit you with stories they have made up. Their condemnation has long been hanging over them, and their destruction has not been sleeping.[2 Peter 2 v1-3]

For such people are not serving our Lord Christ, but their own appetites. By smooth talk and flattery they deceive the minds of naive people.[Romans 16 v18]

Then we will no longer be infants, tossed back and forth by the waves, and blown here and there by every wind of teaching and by the cunning and craftiness of men in their deceitful scheming.[Ephesians 4 v14]

Do not be deceived: God cannot be mocked. A man reaps what he sows.[Galatians 6 v7]

As I urged you when I went into Macedonia, stay there in Ephesus so that you may command certain men not to teach false doctrines any longer nor to devote themselves to myths and endless genealogies. These promote controversies rather than God's work - which is by faith.[1 Timothy 1 v3-4]

The deceiver of the world

He threw him into the Abyss, and locked and sealed it over him, *to keep him from deceiving the nations any more* until the thousand years were ended. After that, he must be set free for a short time.[Revelation 20 v3]

The great dragon was hurled down - that ancient serpent called the devil or Satan, *who leads the whole world astray.* He was hurled to the earth, and his angels with him.[Revelation 12 v9]

We know that we are children of God, and that *the whole world is under the control of the evil one.*[1 John 5 v19]

BIBLIOGRAPHY

[References marked thus*are particularly recommended]

- 1 Himmelfarb, G. *Darwin and the Darwinian Revolution*, Chatto & Windus 1959.
- 2 Darwin, C. *Life and Letters Vol 1* (Ed. F. Darwin), Murray 1887.
- 3 Ibid vol 2.
- 4 Ibid vol 3.
- 5 Darwin, C. *More Letters Vol 1* (Ed. F. Darwin), Murray 1903.
- 6 Ibid vol 2.
- 7 Lyell, C. *Life* vol 1 (Ed. Mrs. Lyell), John Murray 1881.
- 8 Ibid vol 2.
- 9 North, F.J. *Sir Charles Lyell*, Arthur Barker Ltd. London 1965.
- 10 Fenton, C.L. & M.A. *Giants of Geology*, Doubleday New York 1956.
- 11 Clark, E.D. *Darwin; before and after*, Paternoster Press 1972 (2nd Impression).
- 12 Bibby, C. *Scientist Extraordinary* (T.H. Huxley) Pergamon 1972.
- 13 Huxley, T.H. *Life and Letters* Vol I (Ed. L. Huxley) Macmillan 1903.
- 14 Ibid Vol II
- 15 Ibid Vol III
- 16 Irvine, W. *Apes, Angels and Victorians*, Weidenfeld & Nicolson 1956.
- 17 *Encyclopaedia Brittanica*.
- 18 King-Hele, D. *Doctor of Revolution* (Erasmus Darwin) Faber 1977.
- 19 Darwin, C. *The Origin of Species*, J.M. Dent & Sons 1972.
- 20 Lyell, C. *Principles of Geology* vol 1, Murray 1867.
- 21 Ibid vol 2.
- 22 Patterson, C. *Evolution*, Routledge & Kegan Paul in association with the British Museum (Natural History) 1978.
- 23 Fisher, R.A. "Has Mendel's work been rediscovered?", *Annals of Science* vol 1 no.2.
- 24 Malthus, T.R. *Principles of Population*, Reprint of 6th (Final) Edition-1826, Ward Lock & Co. 1890.
- 25 Gould, S.J. "Catastrophes & Steady State Earth", *Natural History* vol LXXX no.2 February 1975.
- 26 Barlow, N. *The Autobiography of Charles Darwin*, Collins 1958.
- 27 Wilson, L.G. *Charles Lyell. The Years to 1841*, Yale University Press 1972.
- 28 Dewar, D. *The Transformist Illusion*, Dehoff Publications, Tennessee, 1957.
- 29 Nordenskiold, E. *The History of Biology*, Tudor Publishing Co., New York 1928.
- 30 Lewis, C.S. *Mere Christianity*, Collins Fontana Books 1956.
- 31 Polanyi, M. "From Copernicus to Einstein", *Encounter* vol V no.3 pp54-63.
- 32 Lewis, C.S. *Miracles*, Collins Fontana Books 1966.
- 33 *The Dark Side of the Moon*, Faber 1946.
- 34 Hardin, G. *Nature and Man's Fate*, Mentor 1961.
- 35 Tozer, A.W. *The Knowledge of the Holy*, Send The Light Trust 1976.
- 36 Webster, N. *The French Revolution* The Britons Publishing Company
- 37 Macbeth, N. *Darwin Retried*, Garnstone Press London 1974.
- 38 Jones, A. "The Genetic Integrity of the Kinds", Creation Science Movement Pamphlet no.227 July 1981.
- 39 References on Cortical Inheritance:

    a) Klug, S.H. "Cortical studies on Glaucoma" Journal of Protozoology, 15 p321-327 1968.

    b) Nanney, D.L. "Cortical patterns in cellular morphogenesis", Science 160 p496-502 1968.

    c) Sonneborn, T.M. "The differentiation of cells" Proceedings of the National Academy of Sciences of the USA, 51 p915-929 1964.

    d) Sonneborn, T.M. "The evolutionary integration of the genetic material into genetic systems" *Heritage from Mendel* (Ed. R.A. Brink) p375-401 University of Wisconsin Press 1967.

    e) Sonneborn, T.M. "Gene action in development" Proceedings of the Royal Society B, 176 p347 1970.

f) Beisson, J. and Sonneborn, T.M. "Cytoplasmic inheritance of the organization of the cell cortex in Paramecium aurelia" Proceedings of the National Academy of Sciences of the USA, 53 p275-282 1965.

g) Tartar, V. "Morphogenesis in Protozoa" Research in Protozoology, 2 p1-116 1967.

h) Willie, J.J. "Induction of altered patterns of cortical morphogenesis and inheritance in Paramecium aurelia" Journal of experimental Zoology, 163 p191-213 1966.

40   Schaeffer, F. *Escape from Reason* Inter-Varsity Press 1968

41   *Buzz* Magazine, April 1979 p25.

42   Burke, E. *Reflections on the French Revolution*, J.M. Dent & Sons 1951.

43   Raverat, G. *Period Piece*, Faber & Faber 1954.

· 44   Archer, G.L. *A Survey of Old Testament Introduction*, Moody 1972.

45   Bowden, M. *Ape-Men — Fact or Fallacy?* (1st Ed.), Sovereign Publications 1977.

46   Bowden, M. *Ape-Men — Fact or Fallacy?* (2nd enlarged Ed.), Sovereign Publications 1981.

47   Morris, H.M. *The Troubled Waters of Evolution*, Creation Life Publishers, San Diego 1980.

48   Pickering, H. *Chief Men among the Brethren*, Pickering & Inglis.

49   Coad, F.R. *A History of the Brethren Movement* (2nd Ed.), Paternoster 1976.

50   Bibby, C. *T.H. Huxley*, Watts, London 1959.

51   Wallace, A.R. *My Life* vol 1, Chapman and Hall 1905.

52   Ibid vol 2.

53   Rusch, W.H. "Ontogony Recapitulates Phylogony", Creation Research Society Annual (June) vol 6 no.1 1969.

54   De Beer, G. *An Atlas of Evolution* Nelson 1964.

55   Popper, K. *"Conjectures and Refutations"* Routledge & Kegan Paul 1963.

· 56   Courville, D.A. *The Exodus Problem and its Ramifications* (2 vols), Challenge Book, California 1971.

· 57   Velikovsky, I. *Ages in Chaos*, Sphere Books 1973.

58   Ospovat, D. "Lyell's theory of climate" *Journal of the History of Biology* vol 10 no.2 p317-339 Fall 1977.

59   Thompson, W.R. Introduction to *The Origin of Species* Everyman Library no. 811 Dent 1956.

60   Herbert, S. "The place of Man in the Develoment of Darwin's Theory of Transmutation" part II, *Journal of the History of Biology* vol 10 no.2 p155-227 Fall 1977.

61   Wrangham, R. "The Bishop of Oxford" *New Scientist* p450 9 August 1979.

62   Gruber, H.E. *Darwin on Man* (2nd Ed.) University of Chicago Press 1981.

63   Feuer, L. "Is the Darwin-Marx correspondence Authentic?" *Annals of Science*, 32 p1-12 1975.

64   Feuer, Colp et.al. "On the Darwin-Marx correspondence" *Annals of Science*, 33 no.4 July 1976 p383-394.

65   Henbest, N. " 'Oldest Cells' are only weathered crystals" *New Scientist* vol 92 no.1275 p164 15 October 1981.

66   Gould, S.J. "Punctuated equilibria; the tempo and mode of evolution reconsidered" *Palaeobiology* vol 3 p115-151 Spring 1977.

67   Lucas, J.R. "Wilberforce no ape" (Letter) *Nature* vol 287 p480 9 October 1980.

68   Crabtree, H. (Letter) *Nature* vol 289 p344 29 January 1981.

· 69   Steidl, P.M. *The Earth, the Stars and the Bible* Presbyterian and Reformed Publishing Co. 1979.

70   Woodmorappe, J. "Radiometric Geochronology Reappraised" Creation Research Science Quarterly Sept. 1979 v16 n2 p102.

71   Lubkin, G. *Physics Today* v32 n17 1979.

72   Haldane, J.B.S. *Possible Worlds* 1927

73   Huxley, J. *Evolution as a Process*

INDEX

By the same author

APE-MEN – Fact *or Fallacy?*

A critical examination of the evidence highlights the very speculative theories based upon inadequate fossil evidence, and reveals the very dubious circumstances surrounding their discovery.

Summary of Contents

PILTDOWN. The considerable body of little publicized evidence which incriminates Teilhard de Chardin S.J. *Professor Douglas' accusation. *The involvement of the British Natural History Museum in the planning and execution of the hoax.*

APE–MEN 'EVIDENCE'. The very speculative nature of the evidence for 'ape-men', and the presumptuous way in which this is presented.

EARLY HOMO SAPIENS. Their existence in deeper strata than those of 'ape-men'. The superficial reasons given by the experts for their rejection.

PEKIN MAN. A 25 ft. high ash heap, bone tools and other evidence of human habitation of the site virtually suppressed by the experts in China. *The appearance—and rapid disappearance in 13 days—of ten skeletons.* Details of the later discovery of further human skeletons delayed for five years. Ape-like skulls reconstructed with human features. Investigation of the disappearance of the fossils at the time of Pearl Harbour suggests that they were found by the Japanese and passed to the Americans after the war, only to disappear again.

JAVA MAN. Dubois' concealment of human skulls for thirty years. The faking of scientific illustrations by Dubois' supporter, Professor Haeckel. The strange circumstances of the discovery of further fossil 'evidence' of Java man.

NEANDERTHAL MAN. The evidence that these were true men suffering from rickets, arthritis and syphilis.

THE AFRICAN APE-'MEN'. The admission by several experts that all these fossils are simply apes with no real human features.

OLDUVAI GORGE. L. S. B. Leakey's discoveries examined.

EAST RUDOLF. Richard Leakey's '1470' man really a human skull. How this 'awkward' fossil was quietly 'buried'.

HADAR (Ethiopia). D. C. Johanson's meagre collection of fossils shown to be only those of apes.

LAETOLIL. Mary Leakey's discovery of '3.6 million year old' footprints are clearly those of human beings. The evidence of similar tracks with those of dinosaurs in America.

CONCLUSION. The inadequacy of the fossil evidence for ape-men. *How the scientific establishment suppressed the publication of un-welcome evidence. *The basic motive for the belief in the ape-men theories.*

(*2nd Edition)

SOVEREIGN PUBLICATIONS

BOX 88, BROMLEY, KENT, BR2 9PF.

APE-MEN – Fact *or Fallacy?*

REVIEWS OF THE FIRST EDITION

English Churchman. 'This is a most learned, factual and highly-documented treatise on the subject of the findings, during the last two centuries, of certain ape-like fossil fragments from which scientists have deduced that man is descended directly from the apes.

The author exposes with pitiless logic and documentation the "last-word" theories of scientists, geologists and anthropologists, consequent upon the discoveries of such things as the "Piltdown Man", exposing many of them as frauds and hoaxes—which in the case of the Piltdown findings is now universally admitted.

This is a book of absorbing interest, especially to Christian teachers of Science and R.E., since any Secondary schoolboy will tell you glibly that "Man is descended from monkeys—Science has proved it." Mr. Bowden's exposures are quite unanswerable, . . .'

International Catholic Priests Association. 'This is one of the most important works for years on the ape-men fossils, and it shines a bright light on four aspects. Firstly, the author shows that the ape-men fossils are dubious in the extreme. Secondly, he shows that evolutionists have concealed or minimized fossils of real men as ancient as these of their supposed ancestors, the ape-men. Thirdly, the ape-men have not been "discovered" by a huge army of scientists, but rather by a tiny group, numbered almost in single figures, travelling from hoax to hoax. Lastly, many will conclude from this work that right in the centre of this group was none other than Teilhard de Chardin . . . This book is written by a clear thinker with a scientific approach who has long studied the original books and papers, weighed one account against another, and has now given us the results in a condensed yet clear way. Everyone should have this book and make sure that their public library also has it.'

Evangelical Times. 'In places it reads like a detective story . . . Indeed one is left with the inescapable impression that the trail of suspicion goes beyond Piltdown . . . [a book] which will be worth adding to your collection.'

Evangelical Action (Australia). . . . Although written in a scholarly and technical manner, "Ape-men, Fact or Fallacy" is, nevertheless, quite easily understood by the non-technical reader and I recommend it to all those who, like myself, have to provide answers to evolutionary questions.'

Prophetic Witness (Review by Dr. F. A. Tatford). '. . . This is an important book, covering the whole ground of fossil evidence for evolution, and it cannot be ignored. We commend it to our readers.'

Fellowship (Review by Duane T. Gish, Ph.D.—Associate Director, Institute for Creation Research, California). 'Anyone interested in the fossil evidence for the ancestry of man should have a copy of Malcolm Bowden's book . . .'

SECOND ENLARGED EDITION
(60 extra pages)
267 pages. 65 Illustrations.
8⅜" x 5⅜". Fully referenced.
Index. 4 colour cover.
ISBN 0950604216